Non-Operated Petroleum Assets

Fabrice Davy OKASSA

Non-Operated Petroleum Assets

Foreword of His Excellency Haitham Al-Ghais

L'Harmattan
Congo Brazzaville

219, Avenue Nelson Mandela; Brazzaville

http:// www.harmattan-congo.com

ISBN : 978-2-336-57416-5
EAN : 9782336574165

To my late father — keeper of numbers, herald of free thought: you were, you are, and shall for ever be an indomitable soul.

Foreword

The world today operates within an increasingly complex landscape. Development challenges persist across the globe, while environmental issues remain unresolved and demand urgent attention. Equally pressing — if not more so — is the reality that millions of people still live in energy poverty, a condition that impacts daily life, endangers health, and prevents individuals and communities from achieving a dignified standard of living.

In the energy sector, we are all too familiar with the multifaceted challenges faced by traditional energy sources, particularly oil and gas. These challenges often stem from criticisms not grounded in a full understanding of the realities at hand—such as calls for divestment, sharp accusations concerning greenhouse gas emissions, and, perhaps most concerning, demands to abandon the use of sovereign natural resources that countries have been endowed with and over which they hold recognized authority.

Against this backdrop, the publication of *Non-Operated Oil Assets (NOPAs)*, authored by respected oil industry expert Mr. Fabrice Davy Okassa, is both timely and essential. Drawing on decades of experience in the oil and gas sector as well as in policy circles at both national and international levels, Mr. Okassa offers a thorough and thoughtful contribution to the field. This academic and professional work addresses the key aspects of oil asset operations, with a distinctive focus on a critical—but often overlooked—perspective: that of the non-operating party.

Unlike the operator who manages day-to-day field activities, the non-operating party plays a behind-the-scenes yet pivotal role. Mr. Okassa begins by clearly defining the roles and responsibilities that differentiate the non-operator from the operator. His research makes a compelling case that non-operating partners are fundamental to the success and viability of oil projects. To safeguard their rights and mitigate their exposure to operational risks, a robust governance framework is required—one that adheres to specific principles thoroughly explored in the book.

The work reaches its core in its examination of rational management foundations designed to protect the non-operator's interests. It presents mechanisms for overseeing operations, incorporating international best practices and modern technologies. Financial and accounting dimensions are also rigorously addressed, with a detailed breakdown of the key concepts relevant to non-operating stakeholders.

Throughout, the author underscores the importance of collaboration and communication between operators and non-operators. As with any joint venture, trust is essential. Only through strong partnerships can projects deliver optimal outcomes—benefiting both the consortium and the host nation.

This publication arrives at a pivotal moment for the energy sector. As environmental concerns, digital transformation, and fiscal discipline reshape global investment strategies, non-operating positions have emerged as a vital tool for building portfolio resilience—enabling diversification, enhancing local content, and optimizing fiscal outcomes.

For resource-owning nations, especially in the Global South, this book serves as a practical guide to maximizing national benefits from hydrocarbons through active participation, oversight, and alignment with broader development goals. For private companies and institutional investors, it offers actionable insights into managing financial risk, navigating complex contractual structures, and ensuring high performance in joint ventures.

Non-Operated Oil Assets fills a crucial gap in both literature and industry practice. By blending technical depth with institutional insight, Mr. Okassa has created a resource for energy professionals, policymakers, investors, and scholars seeking to navigate the future of shared resource management in an era defined by complexity and accountability.

His Excellency Haitham Al-Ghais
Secretary General of the Organization of the

Petroleum Exporting Countries (OPEC)

Introduction

This English edition is a translation of the French original *Actifs pétroliers non opérés*. It includes an additional Chapter 11, "Economic Evaluation of Petroleum Assets," added specifically to this edition to provide a concrete application framework; this chapter is not present in the French version.

This book explores a strategic dimension of the hydrocarbon industry: NOPAs. Often overlooked by the public, they remain central to both economic and institutional stakeholders. These assets are defined as participations in oil projects where the investor does not bear direct responsibility for technical or day-to-day operations. Although seemingly passive, this investment model serves as a lever for diversification and risk reduction for companies engaged in a sector characterized by massive upfront costs and significant technical uncertainties. Collaboration between operators and non-operating partners lies at the heart of this dynamic, enabling the sharing of costs and responsibilities to maximize returns.

Companies holding NOPAs enjoy a unique form of strategic flexibility. This allows them to diversify their portfolios by entering multiple geographic or technological markets while limiting exposure to any single project. For example, a company may invest in deepwater offshore assets in the Gulf of Mexico while also participating in conventional projects in Africa or Asia. Such positioning strengthens resilience in the face of market fluctuations and geopolitical uncertainties.

The illustrative examples detailed in this book, such as Angola's Block 17 or Brazilian offshore projects, show how effective collaboration enables the pooling of resources and expertise. These cases also highlight the importance for non-operating partners to ensure the competence and reliability of the lead operator, since the success of NOPAs largely depends on the latter's performance. Consequently, rigorous oversight—often exercised through management committees—is indispensable to guarantee transparency and compliance in operations.

The book also emphasizes the revenues generated by such assets. In the form of royalties, dividends, or production shares, these financial flows enable investors to secure passive profits while minimizing operational involvement. Nonetheless, risks remain present—technical, financial, and environmental. Industrial accidents, oil price volatility, and environmental incidents, such as the Deepwater Horizon explosion in 2010, illustrate potential consequences. These risks underscore the necessity for investors to conduct thorough due diligence before committing capital.

As for regulatory frameworks, petroleum agreements and joint-venture contracts play a crucial role in defining the rights and obligations of each party. These legal instruments ensure transparent management aligned with the strategic objectives of both States and investors. International initiatives, such as the Extractive Industries Transparency Initiative (EITI), are also referenced as benchmarks for promoting responsible and equitable governance.

For States—particularly those whose economies depend heavily on oil revenues—the management of NOPAs must be treated as a strategic lever. Adapted fiscal policies, strict oversight, and transparency mechanisms are essential to maximize economic benefits while protecting the environment and addressing the expectations of local populations. Continuous improvement of national capacities—through training of professionals and the adoption of advanced technologies—is equally fundamental.

In summary, this book provides a comprehensive and in-depth perspective on NOPAs. It sheds light on their strategic role in portfolio diversification, risk sharing, and the optimization of natural resources. The examples, analyses, and recommendations offered make it a valuable tool for policymakers, investors, and operators in the petroleum industry.

The Author

Okassa Fabrice Davy

Chapter 1

Understanding NOPAs

The oil industry is a highly complex sector, characterized by colossal investments and significant risk-taking. In this context, the use of NOPAs has emerged as an effective solution to enable the sharing of financial, technical, and environmental risks among various stakeholders. NOPAs represent equity interests in exploration, development, or production projects whose day-to-day management is entrusted to a specialized operator. This collaborative model allows for the multiplication of investment opportunities while preserving the specific expertise and skills of each party.

This chapter aims to provide a detailed presentation of what NOPAs are, by examining their definition, specific features, advantages and disadvantages, as well as the roles and responsibilities of the different actors involved. It also highlights the contractual and regulatory frameworks governing their management and analyzes the strategic stakes for States that depend on oil revenues. By drawing on concrete examples from strategic regions such as the Gulf of Mexico, Africa, and Brazil, we will demonstrate how this model is embedded in a logic of sharing and cooperation—indispensable when facing the contemporary challenges of the oil industry.

1.1 Definition and Specific Features of NOPAs

NOPAs are defined as financial interests in oil projects in which the investor does not assume responsibility for the operational management of the project. In other words, a company or investor may hold a stake in an exploration or production license without being responsible for field implementation. Operational management—including exploration, drilling, infrastructure installation, and production—is delegated to a specialized operator with the necessary technical expertise and logistical capabilities.

1.1.1 Sharing Financial and Technical Risks

One of the fundamental principles underpinning the use of NOPAs is risk sharing. In the oil sector, exploration and development projects require massive upfront investments, often amounting to several billion dollars—especially in deepwater operations or complex environments. To avoid bearing the full burden of these costs and inherent risks alone, companies opt to distribute investments among several partners. For example, if a company holds 30% of a project, it bears 30% of the costs and, in return, receives a share—depending on the type of production-sharing contract—of the revenues generated.

This pooling of risks allows each stakeholder to invest in ambitious projects without the threat that bankruptcy or a technical incident could jeopardize their entire portfolio. Moreover, diversification of investment portfolios—commonly adopted by major oil companies such as ExxonMobil, Chevron, TotalEnergies, or Eni—helps stabilize revenues and mitigate the impact of setbacks in a given region. Thus, the same company may be associated with projects in the Gulf of Mexico, West Africa, and Southeast Asia, ensuring a more balanced and resilient revenue stream in the face of market fluctuations.

1.1.2 Flexibility and Focus on Core Competencies

Another major advantage of NOPAs lies in the ability they give companies to focus on their core competencies. In a sector where expertise is often highly specialized—whether in drilling technology, infrastructure management, or refining—some companies prefer to invest in operational projects without assuming day-to-day management responsibilities. This strategy allows them to reap the benefits of a project without having to deal with operational complexity, entrusting it instead to partners with advanced know-how.

Investment funds or technology-driven players, for example, can participate in deepwater exploration projects by relying on experienced operators. This model not only fosters operational efficiency but also encourages skill transfer and collaboration among companies with complementary profiles, thereby strengthening the overall project.

1.1.3 Passive Revenues and Diversification of Income Streams

From a financial standpoint, NOPAs often represent a source of passive income for investors. These revenues typically take the form of:

- **Royalties**: Paid as a percentage of extracted volumes, royalties provide shareholders with steady income regardless of fluctuations in operating costs.

- **Dividends**: Depending on the profits generated by the project, non-operating partners may receive dividends proportional to their equity share.

- **Production Share (Profit Oil)**: In some contracts, a direct portion of production is allocated to non-operating partners, allowing them to benefit from operational performance without direct involvement in the process.

This revenue structure offers investors an attractive diversification strategy, enabling them to capture the benefits of a project while avoiding exposure to daily operational risks.

1.1.4 Risks and Challenges of NOPAs

While NOPAs offer undeniable advantages, they are not free from challenges. One of the main drawbacks lies in the loss of direct control over project management. Non-operating partners must rely on the operator to successfully execute all stages, from exploration to production. This dependency can sometimes lead to strategic disagreements or delays in project implementation. For example, if the operator decides to adjust the development schedule for internal or financial reasons, it may negatively impact the revenues of non-operating partners.

Furthermore, technical risks—such as equipment failures, drilling errors, or industrial accidents—can result in significant cost overruns. Oil price fluctuations also represent a major financial risk. In 2020, a sharp drop in oil prices demonstrated how such variations can render certain projects unprofitable. Finally, environmental concerns, exacerbated by incidents such as the Deepwater Horizon disaster, highlight the need for stakeholders to commit to rigorous monitoring and adherence to strict safety standards.

1.2 Roles and Responsibilities of Stakeholders

NOPAs involve a wide range of actors whose contributions and responsibilities vary depending on their position in the value chain. Close coordination and a clear allocation of roles are essential to ensure the success of an oil project.

1.2.1 The Operator: Pillar of Operational Management

The operator is the central player in managing an oil asset. It assumes full operational responsibility, from initial planning to drilling execution and production management. Its main tasks include:

- **Exploration and Reservoir Evaluation**: The operator carries out seismic surveys, maps subsurface structures, and assesses reservoir production potential. This step is crucial to determine the project's viability.

- **Planning and Drilling Execution**: By mobilizing advanced technologies and specialized equipment, the operator conducts drilling operations while minimizing risks and optimizing yields.

- **Infrastructure Installation and Maintenance**: Construction and management of offshore platforms, pipelines, and processing facilities also fall under its responsibility, ensuring a continuous flow of production.

- **Compliance with Safety and Environmental Standards**: The operator must ensure that all operations comply with safety standards and environmental regulations, conducting regular audits and implementing risk-prevention plans.

- **Communication and Transparency**: The operator is also required to provide detailed and regular reports to non-operating partners, thereby ensuring collaborative and transparent project management.

1.2.2 Non-Operating Partners: Investors and Remote Supervisors

Non-operating partners, although not involved in daily management, play a fundamental role by providing the necessary capital and participating in strategic decision-making. Their contributions materialize through several mechanisms:

- **Financial Contribution and Risk Sharing**: By investing in the project, they share the costs and risks associated with oil exploitation. This pooling allows all partners to limit their individual exposure to unforeseen events.

- **Oversight and Control**: Through management committees and regular audits, non-operating partners exercise oversight, ensuring that the operator complies with contractual terms and established standards. This supervisory role is vital to guarantee transparency and effective project management.

- **Participation in Strategic Decisions**: In some cases, especially when investments are significant, non-operating partners take part in defining the project's overall direction. Their strategic input may relate to the selection of innovative technologies or the adoption of new management methods that optimize the project's overall performance.

1.2.3 Public Authorities and Regulatory Bodies: Guardians of the Public Interest

States, through their public administrations and regulatory agencies, occupy a central position in the management of NOPAs. Their mission is to safeguard national interests, ensure compliance with fiscal and environmental obligations, and guarantee equitable redistribution of revenues from natural resource exploitation. Their actions include:

- **Compliance Oversight**: Authorities ensure that operators and partners comply with health, safety, and environmental regulations, thereby guaranteeing responsible operations.

- **Revenue Management**: The State ensures that revenues from oil projects (royalties, taxes, dividends) are properly collected and redistributed, with a focus on transparency in fiscal mechanisms.

- **Protection of Strategic Interests**: Through the negotiation of contracts and petroleum agreements, governments safeguard sovereignty over natural resources and ensure that projects deliver maximum benefits to the national economy.

- **Promotion of Sustainable Development Initiatives**: By working with local stakeholders and involving civil society, they encourage practices that balance economic performance with environmental protection, thereby contributing to more sustainable exploitation of oil resources.

1.3 Regulatory and Contractual Frameworks

The regulation of NOPAs relies on a set of legal and contractual instruments that clearly define the rights, obligations, and responsibilities of stakeholders. These frameworks secure investments, ensure transparent management, and minimize risks associated with oil operations.

1.3.1 Petroleum Agreements

Petroleum agreements form the legal foundation governing relationships between the State and private operators. They typically specify:

- **Production-Sharing Terms**: Production Sharing Contracts (PSCs) define how production is split between the State and companies, after deduction of operating costs. This arrangement enables the State to recover its initial investments while earning a share of net production.

- **Fiscal and Customs Regimes**: Agreements outline tax rates, royalties, and other financial contributions applicable to oil projects, thus balancing investor attractiveness with State profitability.
- **Environmental and Social Clauses**: Increasingly, agreements include strict requirements on environmental protection and social responsibility, covering risk management, waste treatment, and contributions to local development projects.

1.3.2 Joint-Venture Agreements

Joint-venture agreements, often signed between several companies, organize the distribution of equity and costs within an oil project. They include essential elements such as:

- **Investment Allocation**: These contracts determine each partner's share in the project, proportional to the investment made and risks assumed.
- **Decision-Making Procedures**: Strategic decisions—especially those involving additional investments or adoption of new technologies—often require unanimous approval or a weighted vote according to equity shares.
- **Transfer and Exit Conditions**: To ensure project stability, agreements provide clauses governing the conditions for exiting or transferring stakes, thereby ensuring continuity in operations despite changes in partnership structure.

1.3.3 National and International Frameworks

Beyond bilateral or multilateral contracts, the management of NOPAs is embedded in a broader regulatory environment, which relies on:

- **Safety and Environmental Standards**: At both national and international levels, stringent standards regulate oil activities. These norms aim to minimize risks to the environment and worker safety, requiring regular inspections and the implementation of emergency procedures.

- **Transparency Requirements**: To combat corruption and ensure responsible resource management, numerous international initiatives—such as the Extractive Industries Transparency Initiative (EITI)—mandate the regular publication of revenues and project performance.

- **Alignment with International Standards**: To foster cooperation and attract foreign investment, States must ensure that their regulatory frameworks align with global best practices. This alignment not only strengthens legal stability for projects but also boosts investor confidence.

1.4 Strategic Stakes for the State in Managing NOPAs

For many resource-rich countries, the management of NOPAs carries paramount strategic importance. Although directly managed by private operators, these assets represent a crucial source of revenue for the State. Their oversight must therefore be both proactive and strictly regulated in order to address several major challenges.

1.4.1 Maximizing Revenues and Optimizing Fiscal Returns

One of the State's top priorities is to maximize revenues generated from the exploitation of petroleum resources. To achieve this, several strategies are implemented :

- **Designing an Adapted Fiscal Framework**: This involves defining tax rates, royalties, and equity participation that are both attractive for investors and profitable for the

State. A well-calibrated fiscal regime ensures recovery of invested costs while guaranteeing the State a fair share of profits.

- **Effective Contract Negotiation**: During the signing of Production Sharing Contracts (PSCs) or joint-ventures, the State must secure favorable clauses, such as cost oil provisions or progressive royalties, which increase its revenues as production scales up.

- **Establishing Monitoring Mechanisms**: To prevent revenue leakage or underreporting, the State must implement robust control systems. Continuous monitoring of operator performance allows for the swift identification of irregularities and timely corrective action.

1.4.2 Oversight, Control, and Strengthening Governance

The oversight of NOPAs is a major responsibility of the State. Despite its lack of direct involvement in day-to-day management, the State must ensure that operators comply with established standards and that revenues are properly remitted. This involves:

- **Creating Specialized Oversight Bodies**: Establishing audit committees or agencies with technical and financial expertise to regularly review project performance. These entities verify contractual compliance and adherence to regulations.

- **Use of Monitoring Technologies**: Digital tools—such as real-time tracking platforms or data analysis systems—provide constant visibility on project progress. These technologies make anomaly detection easier and strengthen transparency.

- **Cooperation with Operational Stakeholders**: Building a close, transparent, and reciprocal partnership with operators fosters trust and efficiency. Regular inspections and review meetings reinforce this relationship and contribute to improved governance.

1.4.3 Proactive Risk Management

The risks associated with oil projects are multiple and diverse. Whether technical, financial, or environmental, they require proactive, well-orchestrated management to safeguard national interests. Measures include:

- **Contingency Planning**: To address potential accidents or unexpected production interruptions, both the State and operators develop crisis-management plans, including emergency procedures, financial reserves, and rapid recovery strategies.

- **Creation of Stabilization Funds**: Given the volatility of international oil prices, some States establish sovereign wealth funds. These funds help smooth revenue fluctuations and ensure long-term financial stability.

- **Implementation of Strict Environmental Standards**: To minimize environmental risks, rigorous regulations govern operations. These standards impose high levels of safety and environmental protection, reducing both the likelihood and impact of accidents.

1.4.4 Strengthening National Capacity and Knowledge Transfer

Effective management of NOPAs relies not only on fiscal and oversight mechanisms but also on continuous reinforcement of national technical and administrative capacities. Key initiatives include:

- **Training and Human Resource Development**: Specialized programs in petroleum taxation, contract management, and technical analysis are vital for equipping public officials with the necessary skills. Establishing centers of excellence and innovation labs strengthens national expertise.

- **Collaboration with International Partners**: Partnerships with international institutions, universities, or technology firms foster knowledge transfer and adoption of advanced technologies, enabling local integration of global best practices.

- **Development of Mentorship Programs**: Preparing a new generation of specialists through mentorship and continuous training ensures a smooth generational transition and sustainable expertise in the petroleum sector.

1.4.5 Transparency and Accountability in Resource Management

Transparency is a cornerstone in the management of petroleum assets. It is essential to build trust among citizens, investors, and international partners. Measures to ensure transparency include:

- **Regular Publication of Reports**: Periodic disclosure of detailed information on revenues, investments, costs, and project performance enhances accountability. Public access to these reports enables citizen oversight.

- **Adoption of International Standards**: Aligning with initiatives such as the Extractive Industries Transparency Initiative (EITI) ensures that disclosed data meet strict standards, promoting openness and integrity in resource management.

- **Civil Society Engagement**: Regular consultation with citizens and involvement of NGOs in project monitoring ensures resource management serves the public interest. Participatory digital platforms reinforce this dialogue and contribute to inclusive governance.

1.5 Illustrative Examples of Projects and Partnerships

To better illustrate the functioning of NOPAs, it is useful to examine concrete examples from different regions. These case studies demonstrate how collaboration between operators and non-operating partners enables the execution of ambitious projects despite significant technical and financial constraints.

1.5.1 The Gulf of Mexico

In the Gulf of Mexico—renowned for its vast hydrocarbon resources—numerous projects rely on NOPAs partnerships. Companies such as Chevron operate multiple platforms while sharing investments and risks with partners such as Shell and BP. This collaborative approach not only distributes the high costs of deepwater drilling but also optimizes operations by pooling the expertise of different stakeholders.

1.5.2 Projects in Africa

In Africa, Angola's Block 17 stands out as an emblematic strategic partnership. In this project, TotalEnergies serves as lead operator, while companies like Equinor and ExxonMobil hold NOPAS stakes. This model allows partners to benefit from a large-scale project without bearing the complexities of daily management. Furthermore, collaboration among these companies fosters skill-sharing and optimizes resource development, while national authorities ensure reinforced oversight.

1.5.3 Deepwater Drilling in Brazil

In Brazil, Petrobras—central to the pre-salt region—regularly collaborates with international partners to carry out deepwater drilling projects. The technical complexity and high costs of these operations make NOPAs participations indispensable. This model not only spreads financial risks but also facilitates access to diverse technologies and expertise, thereby boosting the project's international competitiveness.

NOPAs today represent a strategic lever in the global petroleum industry. By enabling the sharing of financial and technical risks, they offer investors—whether major oil companies, sovereign wealth funds, or institutional investors—the ability to diversify portfolios while focusing on core competencies. Outsourcing operational management to specialized operators further optimizes costs and fosters innovation through complementary partnerships.

Regulatory and contractual frameworks—such as petroleum agreements and joint-venture contracts—form the legal foundation ensuring transparency and stability of such investments. At the same time, public administrations and regulatory agencies play a critical role in ensuring compliance, fiscal collection, and protection of national interests.

For States, proactive management of NOPAs requires revenue maximization, the establishment of strong oversight mechanisms, risk anticipation, and capacity building. Transparency and civil society involvement in project monitoring remain indispensable to guarantee the responsible exploitation of natural resources.

Key Takeaways

Chapter 1 introduces NOPAs, presenting a model where investors hold financial stakes in exploration, development, or production projects without assuming operational management. This model allows companies to diversify portfolios while sharing technical and financial risks inherent to large-scale investments. By delegating management to specialized operators, companies benefit from the technical expertise and logistical capabilities required, while securing revenues in the form of royalties, dividends, or production shares.

The text emphasizes the importance of risk pooling, illustrating how geographic and technological diversification stabilizes revenues against market volatility and geopolitical uncertainties. Furthermore, it highlights the necessity of rigorous monitoring and evaluation mechanisms, enabling non-operating partners to effectively oversee operations despite lacking direct control. This collaborative approach, based on trust between investors and operators, reflects a strategic logic aimed at optimizing profitability and resilience of projects while mitigating the inherent risks of the petroleum industry.

Chapter 2

Fundamental Principles of Rational Management

The rational and sustainable management of petroleum resources rests on a set of essential principles that ensure transparency, accountability, rigorous risk assessment, and optimization of control processes. In an increasingly complex economic and geopolitical environment, these principles are not merely regulatory obligations; they are also strategic levers indispensable for strengthening States' credibility and international investors' confidence. This chapter explores these principles in depth, highlighting their stakes, implementation modalities, and the impacts they generate on the long-term viability of oil-sector operations.

2.1 Global Stakes and Good-Governance Imperatives

The World Oil Outlook (WOO) published by OPEC in 2024 clearly shows that oil and gas will continue to play a major strategic role in the global economy through 2050 and beyond. Contrary to forecasts predicting an imminent demand peak, OPEC foresees a significant increase—on the order of 24%—pushing global consumption to over 120 million barrels per day by 2050. This growth will mainly come from non-OECD countries, reflecting their rapid economic expansion and rising energy needs.

In this context, oil and gas will still account for more than half of the global energy mix in 2050. This projection underscores not only their central role in global energy security but also their strategic importance for the economic stability of many producing States that derive substantial revenues from these resources. Indeed, revenues from the oil and gas industry will continue to constitute a considerable share of producer-country budgets, supporting public infrastructure, education, health and social systems, and diversified economic development projects.

For oil companies, these energy resources remain fundamental assets, generating significant financial returns despite challenges linked to the energy transition and market fluctuations. The OPEC report notes that to meet projected demand, around $17.4 trillion in investments will be needed by 2050, thereby creating major economic opportunities for energy-sector companies.

Moreover, recent technological developments and the industry's growing environmental commitments—particularly in reducing emissions via carbon capture, utilization and storage (CCUS) and eliminating routine flaring by 2030—reinforce this essential industry's long-term viability and sustainability. Consequently, principles of transparency are guarantees of durability and financial stability.

2.2 Transparency and Accountability

2.2.1 Transparency: The Foundation of Ethical Governance

Transparency is the cornerstone of the rational management of natural resources. It creates a climate of trust that is indispensable between the host country, investors, and international partners. In petroleum resources—treated as sovereign assets in most States—transparency is all the more crucial. While a few exceptions exist (e.g., in the United States, subsurface resources belong to landowners), most nations pool their resources with public or private partners in order to maximize field development and share petroleum rents fairly.

To achieve this, transparency relies on several essential pillars:

Regular publication of financial and technical reports

Transparent management first requires the regular release of detailed reports. Financial reports must clearly and comprehensively present generated revenues, incurred costs, executed investments, and financial transfers. In parallel, technical reports should provide an overview of reserves status, facility performance, progress in exploration, and technological innovations deployed in the field. Drafted in accessible language and accompanied by concise summaries, these documents allow both sector experts and the general public to understand the operational and financial situation of petroleum activities.

Internal and external control mechanisms

To guarantee full transparency, control mechanisms must be rigorous and complementary. Internal audits conducted by specialized teams continuously verify that operational processes comply with established standards. These controls include financial and operational audits that assess the performance of different production units, identify variances, and propose corrective measures. These internal audits are complemented by external reviews conducted by independent bodies or third-party experts, providing impartial scrutiny and strengthening the credibility of the overall control framework.

Active stakeholder engagement

Transparency is not limited to publishing information; it also entails the active participation of all stakeholders. Regular consultation of citizens, NGOs, investors, and local communities is essential to ensure inclusive and democratic management. Public consultations—through forums, assemblies, or multi-stakeholder dialogue platforms—give concerned parties the opportunity to express opinions, make recommendations, and contribute to defining strategic orientations.

2.2.2 Accountability: Owning Responsibilities

Accountability is intrinsically linked to transparency. It requires public managers and private partners to be fully responsible for their actions and to answer for their decisions to all concerned actors. Implementing accountability mechanisms rests on several axes:

Responding to inquiries and audits

Those responsible for managing petroleum resources must submit to independent inquiries and respond to questions from oversight bodies. This approach helps build trust, as it enables rapid identification of irregularities and transparent remediation.

Presenting measurable results

Each project or program must be evaluated according to precise, measurable indicators—financial, environmental, or social—that track progress against set objectives. Presenting these results in interactive dashboards and audit reports supports proactive management and risk anticipation.

Implementing corrective plans

In the event of failure or non-compliance, corrective action plans must be implemented immediately. These plans should be clearly communicated to stakeholders and be subject to rigorous follow-up to ensure effectiveness. Adopting a continuous-improvement approach is thus a guarantee of reliability and resilience in managing petroleum resources.

By adopting these principles of transparency and accountability, States and their partners not only strengthen the management of their petroleum assets but also consolidate the confidence of citizens and investors. Such an approach helps prevent misconduct and ensures sustainable, ethical, and efficient exploitation of natural resources.

2.3 Assessment of Technical and Financial Risks

Managing petroleum assets takes place in complex environments where technical and financial factors are intertwined. Rigorous risk assessment is essential to ensure operational durability and the long-term profitability of investments. This section addresses the two major categories of risks and presents the tools and strategies used to anticipate and mitigate them.

2.3.1 Technical Risks

Integrity of infrastructure

Pipelines, offshore and onshore platforms, and refining installations are at the heart of operations. Their failure can have catastrophic consequences—operationally and environmentally. It is therefore indispensable to implement regular, rigorous inspection programs, including non-destructive testing, strength tests, and compliance assessments against international standards. Such inspections detect early signs of degradation and enable intervention before a breakdown occurs.

Safety and efficiency of drilling and production procedures

Drilling and production procedures must be designed according to global best practices. Integrating advanced technologies—such as automated well-management systems—reduces the risk of human error, optimizes production, and minimizes unplanned interruptions. Continuous training of operational teams is essential to ensure mastery of modern techniques and the adoption of a robust safety culture. Embedded at every organizational level, this culture forms a bulwark against accidents and contributes to overall operational optimization.

Preventive maintenance and technological innovation

Implementing a preventive-maintenance program extends equipment life and reduces emergency repair costs. Investment in R&D—especially in monitoring technologies and real-time diagnostic systems—provides valuable tools for anticipating failures. Adopting technological innovations such as artificial intelligence and predictive analytics strengthens companies' ability to identify technical risks and deploy appropriate solutions before incidents occur.

2.3.2 Financial Risks

Oil price volatility

One of the most significant financial risks is the volatility of international oil prices. Often driven by geopolitical, economic, or climatic factors, these fluctuations directly affect revenues and project profitability. To counter this destabilizing effect, robust hedging strategies are essential. Using financial instruments such as futures, options, and other derivatives helps limit price-variation impacts on cash flows and supports budgetary stability.

Managing production delays

Delays—arising from technical, logistical, or regulatory issues—are another major risk. They increase operating costs and reduce expected revenues, potentially undermining project viability. To address this, it is crucial to develop precise action plans, implement real-time monitoring mechanisms, and adopt rigorous planning that anticipates potential obstacles.

Preparing for unforeseen costs

Beyond market fluctuations and delays, unforeseen costs—such as emergency repairs, environmental penalties, or sudden increases in input prices—can compromise project profitability. To guard against such events, companies should build safety margins into budgets and establish reserve funds. These financial buffers allow them to absorb unexpected expenses without disrupting overall operations, thereby ensuring sound, proactive financial management.

2.3.3 Forecasting Tools and Mitigation Strategies

Effectively managing technical and financial risks requires advanced modeling tools and appropriate mitigation strategies. Economic simulations occupy a central place among these tools. They make it possible to consider different scenarios and incorporate specific assumptions to identify potential vulnerabilities. These forward-looking analyses provide a solid basis for informed decision-making and facilitate the implementation of targeted corrective measures.

In addition, integrating information technologies—such as real-time data-analytics platforms and algorithmic tools—strengthens companies' ability to continuously monitor operations and respond quickly to unforeseen situations. With these tools, it becomes possible to implement risk-mitigation strategies that combine preventive measures and emergency plans, thereby ensuring greater resilience of oil operations.

2.4 Integration of Key Performance Indicators (KPIs)

2.4.1 The Strategic Value of KPIs in Asset Management

Key Performance Indicators (KPIs) play a decisive role in optimizing the strategic and operational management of oil assets. Integrating them into monitoring and analysis processes makes it possible to assess operational efficiency, identify improvement opportunities, and align with the operator's strategic objectives. KPIs thus provide a holistic view of performance across technical, financial, and environmental dimensions.

2.4.2 Technical KPIs

Technical KPIs are designed to measure operational performance and infrastructure reliability, with an emphasis on optimizing resource recovery. Notable indicators include:

- **Production rate**

Measured in barrels per day (b/d), this KPI evaluates extraction efficiency and reservoir health. Unusual variations may indicate technical malfunctions or changes in facility performance, requiring in-depth analysis to identify and correct root causes.

- **Hydrocarbon recovery factor**

This indicator measures the proportion of recoverable reserves relative to original reserves. Optimizing reservoir management—particularly by using Enhanced Oil Recovery (EOR) techniques—can significantly increase this rate, ensuring more profitable and longer-term field exploitation.

- **Reserve Replacement Ratio (RRR)**

This ratio assesses the operator's ability to replace produced reserves. A rate below 100% indicates reserves are not being replaced at the pace of extraction, which may jeopardize long-term sustainability.

- **Reservoir management and sustainability of recovery**

Applying dynamic reservoir models, combined with seismic monitoring campaigns and fluid-injection techniques, helps optimize hydrocarbon recovery while minimizing the natural decline of wells.

- **High useful uptime**

Operational availability is crucial to avoid unplanned shutdowns. This KPI measures the percentage of time equipment is operational, helping identify recurring failure causes and enabling preventive actions.

- **Asset durability and cost control**

Rigorous tracking of maintenance and renewal investments reduces unforeseen costs and extends infrastructure life. This indicator evaluates the operator's ability to limit emergency repairs and maintain optimal performance.

- **Equipment availability**

Expressed as a percentage of total operating time, this indicator measures maintenance effectiveness and equipment reliability. High availability reflects proactive management and continuous optimization of technical processes.

2.4.3 Financial KPIs

Financial KPIs monitor profitability, optimize investments, and ensure efficient management of financial resources. Essential indicators include:

- **Cost per barrel**

This distinguishes operating expenditures (OPEX) during regular production phases from capital expenditures (CAPEX) for capacity expansion or upgrades. Differentiating CAPEX and OPEX helps target periods where capital spending must be justified by significant returns.

- **Return on Investment (ROI)**

ROI measures the ratio between net gains generated by a project and total costs incurred. Rigorous ROI analysis evaluates current project viability and steers investment decisions toward higher-return initiatives.

- **Clarification of non-productive CAPEX**

Every capital investment should be directly linked to reserve additions or substantial operational performance improvements. Absent such justifications, the operator must clearly explain the necessity of the expenditure to avoid budget drift.

- **Profit margin**

Calculated as the difference between revenues and all costs (fixed and variable), margin is a fundamental indicator of operational profitability. Continuous monitoring detects deviations and supports optimization measures to keep margins aligned with strategic objectives.

2.4.4 Environmental KPIs

Amid the energy transition and increasingly stringent environmental regulation, environmental KPIs are critical. They measure the environmental impact of oil activities and ensure compliance:

- **Compliance with ecological standards**

This indicator evaluates adherence to international environmental protection standards. Discharge management, waste reduction, and biodiversity preservation are key axes to ensure ecosystem-respectful operations.

- **Reduction of greenhouse gas (GHG) emissions**

Limiting the carbon footprint is a strategic priority across the sector. This KPI measures efforts to reduce CO_2, methane, and other GHG emissions—through, for example, adoption of CCUS technologies and energy optimization of industrial processes.

2.4.5 Integrating KPIs into Reporting Systems

KPI effectiveness depends on integration into modern, high-performance reporting systems. Digital tools—such as interactive dashboards and real-time analytics platforms—centralize data from operational units and provide immediate performance visualization. Regular tracking, coupled with internal audits, ensures KPIs remain aligned with strategic goals and quickly surface deviations that could hinder overall performance. Standardized reports and automated analytics further strengthen transparency and the responsiveness of the entire management framework.

2.5 Optimizing Monitoring and Audit Processes

In a sector as complex as oil, optimizing monitoring and audit processes is a strategic pillar to ensure compliance, improve performance, and reinforce stakeholder confidence. These processes—which encompass continuous operational surveillance and periodic evaluations—aim to ensure each link in the production chain meets set standards and that corrective actions are implemented as soon as deviations are detected.

2.5.1 Rigorous Monitoring Processes: A Key to Responsiveness

Effective monitoring relies on continuous observation of key performance indicators. The KPIs presented above enable real-time evaluation of production, operating costs, and facility availability. Proactive surveillance makes it possible to:

- **Detect anomalies quickly**

Real-time monitoring systems immediately flag drifts from preset targets. Such responsiveness is essential for deploying corrective actions before issues escalate.

- **Anticipate operational needs**

Careful data analysis helps forecast necessary adjustments in maintenance, logistics, or workforce organization. This anticipation reduces interruption risks and ensures optimal continuity of operations.

- **Optimize costs and timelines**

Regular tracking of operating costs identifies optimization levers and prevents budget overruns. Likewise, monitoring production timelines helps coordinate activities efficiently, limiting delays and extra costs.

2.5.2 Regular, In-Depth Audits: Strengthening Compliance and Transparency

Audits are indispensable to verify compliance with contractual obligations, regulatory norms, and internal HSE commitments. They include:

- **Contractual audits**

Verify that terms and conditions between operators, partners, and suppliers are respected, checking correct application of contractual clauses and identifying deviations.

- **Regulatory audits**

Conducted by external oversight bodies, these ensure operations comply with applicable laws and regulations. Adherence to local and international standards avoids penalties and disruptions while strengthening the operator's credibility with regulators.

- **Environmental audits**

Focused on environmental impacts, these evaluate discharge management, ecosystem protection, and the effectiveness of emission-reduction measures—key to meeting ecological standards and enhancing corporate reputation.

2.5.3 Importance of Site Visits: Direct, Concrete Assessment

While digital data and automated reports offer a synthesized view, site visits remain essential for concrete assessment of installation conditions. They enable:

- **Direct observation**

On-the-ground presence lets auditors and managers visually assess equipment condition, infrastructure quality, and project progress—validating monitoring data and spotting unrecorded anomalies.

- **Constructive exchanges**

Being on site facilitates direct dialogue with operational teams, clarifies challenges, and identifies improvement paths—enriching data analysis and improving coordination across organizational levels.

- **Safety-practice evaluation**

Site visits also verify the effective implementation of safety and risk-management practices essential to prevent accidents and ensure a healthy work environment.

2.5.4 Integrating Digital Tools: A Lever for Performance

The rise of digital technologies has transformed monitoring and audit management. Integrating digital solutions enables:

- **Data centralization**

ERP systems and analytics platforms centralize critical information in a single space, facilitating decision-making with real-time operational visibility.

- **Predictive analytics**

Advanced modeling algorithms anticipate potential failures and forecast maintenance or process-adjustment needs—cutting downtime and optimizing resource allocation.

- **Traceability and process standardization**

Digital tools automate repetitive tasks and ensure full traceability. Standardized procedures deliver consistent indicator tracking and audit execution, strengthening transparency and data reliability.

2.5.5 A Proactive Approach for Continuous Improvement

Optimizing monitoring and audit processes is not a static endpoint but an evolving, dynamic process. To ensure continuous improvement:

- **Training and skills development**

Regular training for monitoring and audit teams enhances technical and analytical capabilities. Programs focused on technological advances and risk-management best practices improve mastery of digital tools and adaptation to sector changes.

- **Updating systems and methodologies**

In a constantly changing context, monitoring systems and audit methods must be updated regularly to meet operational and regulatory demands—maintaining high effectiveness and responsiveness.

- **Periodic process evaluation**

Embedding periodic evaluations, combined with internal audits, measures process effectiveness and identifies improvement avenues. Regular reviews offer opportunities to recalibrate strategies and optimize resources to better address emerging challenges.

Key Takeaways

Chapter 2 lays out the foundations of rational management of NOPAs, emphasizing a comprehensive approach rooted in transparency and accountability. It underscores the need for internal and external control mechanisms to ensure regular publication of financial and technical reports—indispensable for informed decision-making. The chapter also covers rigorous assessment of technical and financial risks using advanced modeling tools, sensitivity analyses, and forward-looking scenarios to anticipate market fluctuations and prepare appropriate action plans. Integrating KPIs enables measurement of operational, financial, and environmental effectiveness, while adopting modern reporting systems and interactive dashboards facilitates proactive management and continuous process improvement. In sum, the chapter defines a management strategy grounded in the collection, analysis, and control of information—sine qua non conditions for ensuring the stability and sustainability of investments in the oil sector.

Chapter 3
Technical Oversight of NOPAs

Technical oversight of NOPAs is an essential mission for partners who are not directly involved in the day-to-day operations of oil fields. In an environment where risk control and optimization of technical performance are critical, this role spans a wide range of activities—from verifying activity schedules to detailed analysis of technical reports, ensuring compliance with international standards, and securing data management. This chapter details the processes and methodologies put in place to ensure transparent and robust asset management, while guaranteeing that investments and operations are carried out in line with quality and safety standards.

3.1 Monitoring Operator Activities

The first step in technical oversight is to closely monitor the operations conducted by operator partners. This monitoring is crucial to validate that each activity complies with contractual commitments and industry best practices. It rests on several pillars:

3.1.1 Verification of Activity Schedules

Rigorous planning is the cornerstone of any oil operation. Reviewing the programs submitted by the operator ensures that commitments on drilling, maintenance, and production are being met:

- **Program review**

Non-operating partners perform a detailed analysis of activity calendars. They verify the coherence of programs with the project's strategic objectives, ensuring that each phase is properly calibrated to achieve expected results. This includes assessing the operator's ability to meet deadlines and anticipate potential delays.

- **Risk identification and anticipation**

Particular attention is paid to detecting potential risks. Delays, inconsistencies in task sequencing, or technical contingencies can lead to cost overruns or production losses. Preventive analysis identifies these pressure points and enables corrective measures as soon as early signs of dysfunction appear.

- **Prioritization of critical steps**

Certain phases—especially drilling in zones with high geological potential—are deemed critical. Prioritizing these activities optimizes the project's overall economic impact. Partners also ensure that resources allocated to critical steps are adequately planned and that associated KPIs are defined and tracked.

3.1.2 Monitoring Daily Operations Reports

Day-to-day operational control relies on constant, granular analysis of reports produced by the operator. These reports compile real-time operational data and are invaluable for quickly detecting deviations from initial forecasts.

- **Data and indicator analysis**

Each daily report is carefully parsed to extract information on production volumes, any interruptions, technical incidents, and unforeseen events. These data measure operational efficiency and help identify anomalies that could affect overall performance.

- **Comparison with forecasts**

Confronting actual performance with initial forecasts is indispensable to evaluate estimate reliability. In the event of significant variances, in-depth analysis clarifies root causes—be they technical issues, unfavorable geological conditions, or unexpected market shifts.

- **Trend detection and predictive analysis**

Beyond daily control, statistical analysis of reports over several days or weeks reveals emerging trends. Detecting recurring anomalies or unusual variations in operational indicators enables preventive actions and timely adjustments to operational strategies.

3.1.3 On-Site Inspection

To ensure operational compliance and execution quality, regular site inspections are organized. Field visits allow direct assessment of work execution and interaction with technical teams.

- **Scheduled inspections and field visits**

Programmed visits verify that operations are conducted in accordance with approved plans. On-site inspections also provide first-hand information on the condition of facilities and equipment performance.

- **Safety and procedure control**

Strict adherence to safety standards is imperative in the oil industry. Site checks include rigorous verification of compliance with safety protocols, team training, and the condition and maintenance of critical equipment such as blowout preventers (BOP) and other safety devices.

- **Collection of lessons learned**

Gathering feedback from field teams is essential for continuous improvement. Formal and informal exchanges help identify improvement areas, optimize maintenance procedures, and strengthen operational safety.

3.2 Analysis of Technical Reports

Analyzing technical reports is a decisive component of monitoring NOPAs. Produced at various stages of the asset life cycle—exploration, development, or production—these reports provide detailed information on operational performance and viability.

3.2.1 Processing Drilling Logs

Processing drilling logs is the first step in technical evaluation. This meticulous process extracts essential information on subsurface geology and hydrocarbon potential:

- **Identification of geological formations**

Log analysis precisely identifies formations encountered while drilling. These data are crucial to estimate the presence and quality of hydrocarbons, based on parameters such as porosity, permeability, and fluid saturation.

- **Comparison with geological forecasts**

Collected data are systematically compared with pre-established geological forecasts. This comparison refines reservoir models and adjusts exploration strategies. In case of divergence, additional investigations are conducted to understand gaps and revise initial assumptions.

- **Assessment of production potential**

Results from drilling-log analysis provide a first estimate of recoverable volumes. This assessment enables non-operating partners to validate the economic feasibility of future investments and guide strategic decisions on field development.

3.2.2 Analysis of Production Tests

Once drilling is complete, production tests provide a concrete view of the reservoir's actual performance. Analyzing these tests is crucial to identify weaknesses and propose technical improvements:

- **Review of flow rates and pressures**

Data on extraction rates and pressure variations are key indicators of well productivity. Fine-grained analysis can reveal issues such as blockages, pressure losses, or inefficiencies in pumping equipment.

- **Quality control of hydrocarbons**

The composition of produced hydrocarbons is also evaluated. Significant variations in quality may indicate malfunctions in separation processes or issues linked to the geological composition of the formation.

- **Assessment of surface equipment performance**

Beyond reservoir aspects, analysis includes evaluation of surface equipment. The optimal operation of processing, separation, and transport facilities is verified, allowing identification of optimization or replacement opportunities to maximize operational efficiency.

3.2.3 Inspection of Facilities and Operating Units

Installed infrastructure—platforms, subsea facilities, and processing units—must meet strict performance and safety standards:

- **Verification of technical compliance**

Each installation undergoes a series of technical checks to ensure compliance with prevailing standards. This includes visual inspection, functional testing, and verification of system integration.

- **Identification of infrastructure risks**

Facility analysis identifies potential risk areas, such as premature wear of certain equipment or manufacturing defects. These insights are crucial for preventive measures and for planning corrective or preventive maintenance work.

- **Optimization of system integration**

Overall performance depends on effective interconnection of production, surveillance, and safety systems. Regular audits of system integration reduce malfunction risks, improve coordination among operating units, and lower costs related to delays or failures.

3.2.4 Reservoir Modeling and Simulation

Deploying advanced numerical models is indispensable for forecasting future reservoir performance and planning investments:

- **Model construction and validation**

Using data from drilling logs, production tests, and facility checks, reservoir models are built. These models integrate geological and operational parameters to simulate various exploitation scenarios.

- **Simulation of development scenarios**

Models allow multiple scenarios to be considered based on assumptions regarding production volumes, pressure changes, and hydrocarbon price fluctuations. These simulations provide a forward-looking view that helps partners anticipate field evolution and adjust strategy accordingly.

- **Adjustment of investment strategies**

Simulation results are used to optimize investment planning. The goal is to maximize hydrocarbon recovery while avoiding unnecessary or poorly calibrated spending relative to the project's economic outlook.

3.2.5 Benchmarking and Lessons Learned

To ensure optimal competitiveness and performance, results must be compared with recognized industry standards:

- **Technical and operational benchmarks**

Facility and well performance are compared against best practices and standards observed in similar oil fields. This comparative analysis identifies performance gaps and proposes concrete improvement pathways.

- **Integration of lessons learned**

Lessons from past operations—whether observed anomalies or initiatives that delivered performance gains—are integrated into decision-making. Sharing lessons learned among partners strengthens operational resilience and supports a continuous-improvement approach.

- **Evolving standards and adoption of innovation**

In a perpetually evolving technological context, alignment with international benchmarks also requires integrating technical innovations. Technology watch and evaluation of new solutions (advanced monitoring tools, predictive diagnostic systems, etc.) are essential levers for maintaining heightened competitiveness.

3.3 Compliance with International Standards

Compliance with international standards is an essential requirement in the oil sector, both to ensure operational safety and to protect the environment. This oversight is structured around several pillars:

3.3.1 Environmental Standards

The oil sector is subject to strict regulations designed to limit the environmental impact of operations. Technical oversight therefore includes:

- **Compliance with air-emission regulations**

Activities must adhere to standards on greenhouse gas and atmospheric pollutant emissions. Regular checks verify that installations are equipped with effective filtration and emission-reduction systems.

- **Wastewater management and waste treatment**

Treating wastewater and managing industrial waste are critical to prevent contamination of natural ecosystems. Environmental audits validate the conformity of treatment plans and associated infrastructure, as well as the effectiveness of spill-prevention measures.

- **Implementation of spill-prevention plans**

Strict protocols are developed to prevent and manage hydrocarbon spill risks. These plans include rapid-response procedures and containment systems, whose effectiveness is regularly assessed by internal audits and independent inspections.

3.3.2 Safety Standards

Operational safety in the oil sector is a paramount concern that requires rigorous compliance with international standards:

- **Verification of critical equipment**

Safety equipment—particularly blowout preventers (BOPs) and other critical devices—are inspected to ensure compliance with recommendations from recognized bodies such as the **American Petroleum Institute (API)**.

- **Team training and qualifications**

Team competence is ensured through regular training and continuous evaluation. Operators and technicians undergo safety training, including emergency drills, to guarantee effective response in case of incident.

- **Independent safety audits**

Safety reviews conducted by third-party experts validate the adequacy of procedures in place. These audits, often mandated by non-operating partners, result in detailed reports that form the basis for corrective actions and continuous improvement.

3.3.3 Compliance Audits

Compliance audits are a crucial step to verify that operations are carried out in accordance with international standards:

- **Organization of internal and external audits**

Regular audits are conducted by both internal teams and external consultants to verify strict application of safety and environmental standards. These audits highlight potential non-conformities and recommend corrective measures.

• **Preparation of detailed reports and recommendations**

Each audit produces a comprehensive report identifying weaknesses and offering precise recommendations to improve compliance. These reports serve as key tools to guide operational and strategic decisions.

• **Follow-up and implementation of corrective actions**

A rigorous tracking system ensures that audit recommendations are implemented within the required timeframe. Specific performance indicators measure the effectiveness of corrective actions and allow processes to be adjusted if necessary.

3.4 Management of Technical Data and Confidentiality

Secure management of technical data is a fundamental pillar in the oil sector. Information on drilling logs, production reports, and geological assessments is of strategic importance for decision-making.

3.4.1 Centralization and Organization of Data

To ensure optimal use of information, it is imperative to centralize and organize all technical data:

- **Use of secure databases**

Data collected throughout operational phases are stored in secure databases. Advanced data-management solutions ensure traceability and real-time availability of information.

- **Structured organization and facilitated access**

Clear structuring of data—by category, operational period, or equipment type—facilitates access for analysis and audits.

This organization reduces the time required to retrieve and extract critical data.

3.4.2 Controlled Access and Information Security

Confidentiality of technical information is vital to protect the economic and strategic interests of partners:

- **Authentication systems and access rights**

Strong authentication systems, combined with role-based access rights, ensure that only authorized personnel can access sensitive data. These security measures limit risks of leakage or misuse of strategic information.

- **Protection against cyberthreats**

In the face of rising cyberattacks, adopting advanced cybersecurity solutions is indispensable. Firewalls, intrusion-detection systems, and regular backups are deployed to protect data integrity and confidentiality.

3.4.3 Archiving and Long-Term Preservation

Long-term preservation of technical data is another key aspect of managing NOPAs:

- **Secure archiving protocols**

Strict protocols are defined for archiving critical data to ensure accessibility for future needs or regulatory evaluations. These protocols include regular backups and periodic integrity checks of archives.

- **Compliance with legal requirements**

Archiving must comply with applicable legal requirements. Regular audits ensure archives are kept up to date and accessible only to authorized personnel, in line with data-protection regulations.

3.4.4 Integration of New Technologies

Rapid technological evolution offers new opportunities to improve technical-data management:

- **Analytics and artificial intelligence tools**

Integrating advanced analytics tools using artificial intelligence and machine learning makes it possible to detect patterns and anomalies in collected data. These technologies enhance forecasting capabilities, transforming how non-operating partners anticipate and respond to field developments.

- **User interfaces and dashboards**

The development of ergonomic interfaces and interactive dashboards enables decision-makers to quickly access essential information. These tools support fast, informed decisions by providing clear visualizations of operational and technical data.

Key Takeaways

Chapter 3 focuses on the technical oversight of NOPAs. It describes the methods and procedures put in place to ensure that the daily operations of oil projects meet quality and safety standards. It details the importance of regular monitoring of operator activities—through verification of activity schedules, analysis of daily reports, and on-site inspections. It highlights the control tools that allow early detection of anomalies, comparison of actual performance with forecasts, and proactive adjustments. The chapter further explains how analysis of drilling logs, production tests, and reservoir modeling improves understanding of exploitation potential and optimizes interventions. Overall, it demonstrates that technical oversight is essential to ensure the reliability, safety, and performance of NOPAs, while enabling adequate responsiveness to field developments.

Chapter 4

Financial Oversight and Revenue Optimization

Financial management of NOPAs in the oil sector is a strategic discipline that requires deep expertise and advanced analytical tools. This chapter examines methods of financial oversight, cost and revenue analysis, auditing of financial flows, and the use of economic modeling tools to optimize performance and guarantee full transparency in resource management.

4.1 Cost and Revenue Analysis

Rigorous financial analysis forms the foundation for optimal management of NOPAs. It not only makes it possible to assess the profitability of operations but also to identify opportunities for cost reduction and revenue optimization.

4.1.1 Identification of Direct and Indirect Costs

Cost control is essential to ensure the economic viability of an oil project. Expenditures fall into several categories:

- **Capital Expenditures (CAPEX)**

Initial investments in infrastructure development (offshore platforms, pipelines, processing units) represent a significant share of spending. These costs also include:

- Drilling and well-completion expenses, which vary depending on geological complexity and technologies used.

- Acquisition and installation of advanced technologies (monitoring systems, diagnostic tools, separation equipment) that contribute to long-term operational efficiency.

- **Operating Expenditures (OPEX)**

Operating costs encompass day-to-day expenses related to maintenance, logistics, and facility management:

o Maintenance of critical equipment (compressors, pumps, safety systems), which requires close attention to avoid costly breakdowns.

o Logistics expenses, including hydrocarbon transport, which represent a strategic cost item requiring ongoing optimization.

o Costs related to training and qualification of personnel, ensuring optimal skill levels and minimizing operational risk.

- **Royalties and Taxes**

Financial obligations to local or national authorities take the form of royalties and taxes, calculated based on predefined formulas:

o Royalties, often calculated as a percentage of gross revenues, are a fundamental component of the financial model.

o Profit taxes, which vary depending on the fiscal regime of the operating region, must be factored into the project's economic modeling from the outset.

4.1.2 Monitoring and Managing Revenue Flows

Monitoring revenue flows ensures sound and transparent management of cash inflows from asset exploitation:

- **Royalties**

Collection and distribution of royalties must be performed transparently. Control mechanisms ensure payments reflect actual production volumes and adhere to contractual terms.

- **Dividends**

Profit distribution among shareholders depends on a balanced strategy between reinvestment and dividend payout. Effective dividend management maintains investor confidence while ensuring operational sustainability.

- **Production Shares and Production Sharing Contracts (PSCs)**

In production-sharing frameworks, it is vital to rigorously track volumes allocated to each party. Careful management of production shares optimizes revenues and prevents disputes among partners.

4.1.3 Profit Margin Analysis

Analyzing profit margins is a fundamental indicator for assessing operational profitability:

- **Gross Margin**

Calculated before indirect costs, gross margin provides a first measure of operational performance. A high ratio indicates efficient production and effective control of direct costs.

- **Operating Margin**

Including operating expenses, the operating margin provides a broader view of profitability. It highlights areas where optimization is possible to improve overall efficiency.

- **Net Margin**

After deduction of all expenses—including taxes and interest—net margin reflects the true economic value created by operations. This final analysis is indispensable for guiding long-term strategic decisions.

4.1.4 Comparison with Forecasts and Variance Analysis

Comparing actual performance with budget forecasts highlights variances and clarifies their causes:

- **Identification of deviations**

Variances may stem from unforeseen expenses, hydrocarbon market fluctuations, or operational variations. Detailed analysis of these deviations refines initial assumptions and improves forecast accuracy.

- **Use of modern analytical tools**

Interactive dashboards and predictive models are used to monitor key financial KPIs in real time. These tools facilitate continuous performance analysis and help anticipate upcoming trends.

- **Adjustment of financial assumptions**

Based on lessons drawn from observed variances, decision-makers regularly revise financial models to strengthen operational resilience against market uncertainties.

4.2 Auditing Financial Flows

Rigorous financial auditing is indispensable to guarantee transparency and reliability of transactions related to NOPAs. The objective is to verify compliance of financial practices with contractual obligations and detect possible irregularities.

4.2.1 Review of Financial Contracts

Financial contracts—such as Production Sharing Contracts (PSCs) and joint-venture agreements—are examined to ensure their rigorous application:

- **Verification of contractual compliance**

Each financial clause is scrutinized to ensure revenue-sharing terms, royalty calculations, and recovery-cost provisions are respected by all parties.

- **Review of payment terms**

Conditions such as due dates, interest rates, and penalties for late payment are verified to ensure smooth cash flow and avoid disputes.

- **Identification of contractual risks**

Contract analysis identifies ambiguous or potentially conflicting clauses that could lead to future disputes. These risks are then flagged to decision-makers for informed action.

4.2.2 Expense Analysis and Control

Monitoring expenditures made by operators is a central element of financial auditing:

- **Review of financial statements**

Operating costs (OPEX), capital costs (CAPEX), and administrative expenses are compared with budgets to detect anomalies or overruns. This oversight ensures each expense is justified and aligned with expectations.

- **Validation of supporting documents**

Each expense must be supported by documents such as invoices, purchase orders, and contracts. The audit verifies consistency and authenticity to prevent fraud or accounting errors.

- **Control and analysis of variances**

Significant differences between actual expenses and budget estimates are analyzed in depth. Identifying underlying causes allows better forecasting and optimization of future financial management.

4.2.3 Verification of Payments

Rigorous management of inflows and outflows ensures payments are made in line with contractual agreements:

- **Monitoring incoming flows**

Amounts received—whether royalties, production shares, or dividends—are verified in real time. This involves comparing forecasts with actual production data, ensuring complete transparency.

- **Control of payment deadlines**

Payment deadlines are closely monitored to ensure funds are transferred on time. Any delay is immediately flagged and addressed with corrective action.

- **Dispute management and resolution**

In case of irregularities or payment delays, dispute-resolution procedures are implemented to quickly settle conflicts and restore compliance of financial flows.

4.2.4 Production of Audit Reports

Audit results are consolidated into detailed reports that provide the basis for corrective measures and continuous improvement:

- **Identification of irregularities**

Reports highlight areas of non-compliance—such as unjustified expenses or significant budget variances—accompanied by detailed analysis of causes and potential impacts.

- **Recommendations and action plans**

Based on findings, concrete recommendations are made to optimize financial processes and prevent future discrepancies. A detailed action plan is implemented, with tracking indicators to measure corrective-effectiveness.

- **Ongoing follow-up and reassessment**

Audit reports are subject to regular follow-up, ensuring recommendations are applied and financial management remains aligned with contractual and regulatory requirements.

4.3 Use of Economic Modeling Tools

Economic modeling tools play a central role in evaluating and planning the financial performance of NOPAs. They allow decision-makers to anticipate market developments and optimize investments based on robust data and forward-looking analyses.

4.3.1 Discounted Cash Flow (DCF) Models

DCF models are widely used to assess the economic value of an asset by factoring in future cash flows:

- **Calculation of Net Present Value (NPV)**

This indicator quantifies the value of future cash flows discounted at a defined rate, accounting for the cost of capital and the risks associated with operations. NPV is a key criterion for determining the economic viability of a project.

- **Determination of Internal Rate of Return (IRR)**

IRR measures the expected return on investment, enabling comparison with acceptability thresholds and alternative investment opportunities. A high IRR often indicates a well-structured, economically viable project.

- **Formulation and validation of assumptions**

DCF models rely on key assumptions regarding hydrocarbon prices, operating costs, and projected production volumes. These assumptions are regularly reassessed to reflect market fluctuations and lessons learned from operations.

4.3.2 Sensitivity Analyses

Sensitivity analyses measure the impact of variations in key parameters on financial results:

- **Price-variation scenarios**

Oil and gas price fluctuations are simulated to identify critical thresholds and evaluate economic resilience against market volatility.

- **Impact of cost variations**

Changes in operating costs are analyzed to measure their effect on overall profitability. This helps identify cost-reduction levers and anticipate financial stress scenarios.

- **Modeling production-volume fluctuations**

Scenarios incorporating potential variations in production volumes simulate different operational contexts—from optimistic to pessimistic cases. These analyses provide a comprehensive view of associated risks and opportunities.

4.3.3 Economic Scenarios and Strategic Decision-Making

Combining results from DCF models, sensitivity analyses, and economic scenarios allows decision-makers to adopt well-founded investment and management strategies:

- **Development of prospective scenarios**

Optimistic, pessimistic, and neutral scenarios are built to simulate different economic pathways. Each scenario integrates specific assumptions on price evolution, cost levels, and production volumes.

- **Investment prioritization**

By comparing risk-adjusted returns, decision-makers can identify assets offering the best balance between profitability and security. This approach helps concentrate resources on the most promising projects and avoid unprofitable investments.

- **Negotiation and partnership strategies**

Robust data from economic modeling provides a strong foundation for negotiations with partners and investors. It helps justify strategic decisions and enhances the credibility of financial analyses among stakeholders.

4.4 Implementation of Evaluation and Reporting Mechanisms

To ensure optimal and transparent management of NOPAs, it is imperative to establish regular evaluation and reporting mechanisms.

4.4.1 Definition of Key Performance Indicators (KPIs)

Financial and operational KPIs provide a synthesized view of asset performance and support informed decision-making:

- **Production and cost indicators**

Specific indicators measuring production volumes, operating costs, and profitability are defined and tracked in real time. These indicators allow rapid identification of variances from forecasts and prompt corrective actions.

- **Profitability and margin measurement**

Gross, operating, and net margins are calculated regularly to assess overall economic performance. These indicators are benchmarked against industry standards to position performance relative to market norms.

4.4.2 Visualization Tools and Financial Dashboards

Interactive dashboards facilitate communication of financial information to stakeholders:

- **Intuitive visual interfaces**

Data visualization tools—such as graphs and interactive indicators—enable real-time monitoring of KPIs and quick identification of trends or anomalies.

- **Periodic reports and performance reviews**

Quarterly or annual reports consolidate financial and operational results, providing decision-makers and investors with a clear overview. These performance reviews serve as a basis for strategy adjustment and improved asset management.

4.4.3 Coordination and Communication among Partners

Transparency and cooperation among joint-venture partners are essential for long-term success:

- **Regular coordination meetings**

Periodic meetings among partners—including technical and financial teams—allow discussion of operational status, identification of improvement areas, and planning of corrective actions. These meetings also promote sharing of best practices and lessons learned.

- **Collaborative portals and reporting platforms**

Dedicated digital platforms centralize data, facilitate exchanges, and ensure total transparency among stakeholders. These portals provide secure access to critical information and allow real-time monitoring of KPIs.

- **Feedback mechanisms and continuous improvement**

Feedback systems allow teams to share observations and contribute to process improvement. This collaborative approach fosters a culture of continuous improvement, essential for remaining competitive in a constantly evolving sector.

Key Takeaways

Chapter 4 addresses the financial management of NOPAs, highlighting essential methods to analyze and optimize revenues from such investments. It first stresses the importance of rigorous analysis of direct and indirect costs—whether capital expenditures (CAPEX) or operating expenditures (OPEX)—and insists on the careful tracking of financial flows from royalties, dividends, and production shares. It then details the use of modern analytical tools, such as interactive dashboards and predictive models, to compare actual performance with forecasts, identify variances, and adjust financial assumptions. The chapter also underscores the need for continuous auditing of transactions and financial contracts to guarantee transparency, contractual compliance, and prompt resolution of anomalies. Finally, it emphasizes the value of economic modeling tools—particularly discounted cash flow analysis and sensitivity studies—for guiding strategic decisions and ensuring efficient resource allocation in a volatile oil market environment.

Chapter 5

Cooperation and Communication with Operators

Effective management of assets in the hydrocarbons sector depends on the quality and depth of collaboration between asset owners (often States or public entities) and private operators responsible for exploitation. Harmonious cooperation makes it possible to combine the operator's technical and commercial expertise with the owner's strategic vision and national interests. In this context, it is essential to establish a structured framework built on trust, transparency, continuous communication, and shared commitment to common objectives.

This chapter details the various aspects of this interaction, emphasizing the building of strong relationships, effective negotiation to maximize national interests, proactive conflict management, and continuous training of stakeholders. It also incorporates concrete examples and practical recommendations that illustrate how to establish and maintain fruitful cooperation.

5.1 Building Trust-Based Relationships

Trust is the foundation of any lasting professional relationship. In hydrocarbons—where financial, environmental, and social stakes are considerable—establishing and maintaining trust between asset owners and operators is imperative. This trust is built through transparent practices, open communication, and alignment on shared values and objectives.

5.1.1 Transparency in Financial and Operational Reporting

Transparency plays a central role in building trust. Asset owners expect operators to provide detailed financial reports, information on operating costs, and asset performance indicators. Such transparency enables rapid identification of budget deviations, verification of optimal resource use, and anticipation of investment needs.

For example, an operator that regularly publishes reports including financial dashboards, preventive and corrective maintenance records, and environmental performance indicators provides stakeholders with a clear view of operational status. Rigorous tracking of spending related to maintenance, technological upgrades, or continuous improvement initiatives serves not only as a guarantee of sound management but also as a lever to build a long-term relationship of trust.

In addition, implementing regular audits conducted by independent third parties can further strengthen transparency. Covering both financial and operational aspects, these audits provide additional assurance that practices comply with international standards and contractual commitments. Thus, transparency is not limited to information sharing but represents a commitment to providing a faithful and verifiable picture of operational activity.

5.1.2 Regular Meetings

To consolidate trust, it is essential to hold regular meetings between stakeholders. These meetings—whether monthly, quarterly, or semi-annual—provide opportunities to share progress, identify obstacles, and discuss future prospects.

For instance, setting up a steering committee with representatives from both parties can be highly beneficial. Topics such as operational performance analysis, KPI review, financial forecast updates, and strategic adjustments can be addressed. These regular exchanges improve mutual understanding, enable adjustments to action plans based on results, and reinforce cohesion around shared objectives.

Furthermore, these meetings provide an opportunity to recognize and highlight initiatives undertaken by each party. For example, acknowledging efforts in energy optimization or implementation of innovative technologies reinforces the sense of collaboration and shared ownership of the project.

5.1.3 Use of Technological Tools

In the digital age, information technologies provide powerful tools for improving cooperation between asset owners and operators. Collaborative platforms allow instant access to real-time data, facilitating project monitoring and informed decision-making.

Online project-management software, for example, centralizes all operational information: schedules, budgets, KPIs, and even contractual documents. Access to these tools enhances transparency, reduces communication delays, and limits risks of misinterpretation. Automated reporting systems can also generate periodic reports detailing project progress and variances from initial objectives.

Moreover, data-visualization technologies (dashboards) provide a dynamic overview of activity, enabling decision-makers to monitor asset performance in real time. These tools not only facilitate information sharing but also improve responsiveness to unforeseen events, ensuring rapid adaptation of operational strategies.

5.1.4 Commitment to Shared Values

For cooperation between asset owners and operators to be truly sustainable, both parties must share common values. Chief among these are commitments to sustainable development, social responsibility, and compliance with environmental standards.

For instance, adopting green technologies to reduce carbon footprints or implementing corporate social responsibility (CSR) programs can strengthen alignment of interests. Joint initiatives in reducing greenhouse gas emissions, responsible waste management, or biodiversity protection demonstrate a shared commitment to combining economic performance with environmental stewardship.

This shared commitment is also reflected in adherence to codes of conduct or ethical charters defining expected behavioral standards. Signing formal agreements covering respect for human rights, anti-corruption measures, and promotion of diversity enhances the credibility of commitments made by both parties and provides a solid foundation for long-term collaboration.

5.1.5 Mechanisms for Effective Cooperation

Beyond communication and formal commitments, effective cooperation relies on robust operational mechanisms. These include establishing clear protocols for conflict resolution, continuous training for stakeholders, and formalizing agreements that clearly define responsibilities, obligations, and expectations.

For example, adopting a mediation or arbitration protocol to resolve disputes ensures disagreements are addressed fairly and swiftly, preventing communication breakdowns or operational disruptions. Similarly, organizing regular training sessions—technical or managerial—ensures teams remain up to date with regulatory and technological developments.

In addition, detailed contracts and strategic action plans provide clarity and seriousness. These documents should specify not only each party's roles and responsibilities but also performance indicators, monitoring mechanisms, and procedures for revising agreements in light of changing economic or technological contexts.

Thus, effective cooperation between asset owners and operators is not limited to ad hoc information exchange. It is embedded in a shared strategic vision supported by clear commitments, efficient technological tools, and governance mechanisms tailored to the challenges of the sector.

5.2 Negotiations to Maximize National Interests

In hydrocarbons, negotiations between asset owners and operators must maximize national economic benefits while ensuring efficient resource exploitation. Such strategic negotiations must be grounded in a precise understanding of national priorities and applicable regulatory frameworks.

5.2.1 Identification of National Priorities

National priorities in hydrocarbons extend well beyond mere extraction and commercialization. They encompass local job creation, infrastructure development, technology transfer, and environmental protection. It is therefore essential that these priorities are clearly identified and embedded in negotiation strategies.

For example, when signing a production contract, the government may require clauses mandating local employment, vocational training programs, or investment in local infrastructure (roads, hospitals, schools). Likewise, strict environmental requirements may be negotiated, such as installing emission-management systems or financing reforestation projects in production areas.

Clearly defining national priorities enables State representatives to firmly defend the country's interests and ensure maximum economic and social benefits.

5.2.2 Participation in Steering Committees

Direct involvement of national representatives in project steering committees and decision-making bodies provides additional safeguards for defending national interests. Such participation ensures continuous monitoring of contractual compliance and allows operational strategies to be adapted to changing local and international contexts.

For instance, a government representative within the steering committee can intervene to realign priorities, propose corrective measures in case of deviation, or validate innovative initiatives consistent with public sustainable-development policies. This active participation guarantees that field decisions reflect the country's strategic ambitions while fostering knowledge transfer and strengthening local capacity.

5.2.3 Alignment with Regulatory Frameworks

Negotiations must systematically be based on prevailing regulatory frameworks, whether national or international. Compliance with local laws and adherence to international standards are essential to ensure the legitimacy and durability of exploitation contracts.

In many hydrocarbon-producing countries, contracts include specific clauses on skill transfer, environmental protection, and equitable benefit-sharing. Rooted in international agreements or recognized governance norms, these provisions secure investments and build trust among partners.

Alignment with these frameworks also strengthens the country's credibility on the international stage, demonstrating its commitment to responsible and sustainable resource exploitation.

5.2.4 In-Depth Economic and Technical Analysis

Effective negotiations require rigorous economic and technical evaluation of operators' proposals. Such analysis helps assess project viability, identify potential risks, and optimize distribution of responsibilities.

A thorough quantitative evaluation includes analysis of planned investments, production forecasts, operating costs, and expected profit margins. On the technical side, the review covers proposed technologies, team expertise, and the operator's capacity to manage emergencies.

Often carried out in collaboration with independent consulting firms, these evaluations equip negotiators with the tools to defend national interests. They also help identify optimization levers—such as introducing more efficient technologies or implementing cost-reduction measures—thereby improving the overall effectiveness of the project.

5.3 Conflict Management and Dispute Resolution

In a sector as complex and interconnected as hydrocarbons, conflicts are inevitable. They may arise from diverging interests, differing interpretations of contractual clauses, or simple communication errors. Proactive conflict management is therefore essential to prevent disputes from compromising operational continuity or damaging the relationship between asset owners and operators.

5.3.1 Establishing Resolution Mechanisms

Setting up dispute-resolution mechanisms is a key element in preventing and managing conflicts. Clear procedures must be defined in advance, outlining the steps to be followed in the event of a dispute. These may include mediation, arbitration, or, as a last resort, recourse to competent courts.

For example, a contract may establish a conflict-resolution committee composed of representatives from both parties and independent experts. This committee, tasked with amicably reviewing disputes, can intervene quickly to defuse tensions and propose compromise solutions. Similarly, including arbitration clauses ensures impartial resolution if mediation fails.

5.3.2 Promoting a Collaborative Approach

Beyond formal mechanisms, it is crucial to foster a culture of collaboration and dialogue. A collaborative approach relies on the parties' ability to listen, understand, and work together toward balanced solutions.

Follow-up meetings, for instance, can be opportunities to clarify misunderstandings and openly discuss encountered obstacles. By encouraging communication and information sharing, stakeholders can prevent small disagreements from escalating into major conflicts. In this context, training in interpersonal communication and conflict management can be particularly useful for improving daily interactions.

5.3.3 Rigorous Documentation

Keeping comprehensive records of all agreements, amendments, and correspondence related to operations is essential for preventing and resolving conflicts. Rigorous documentation provides concrete evidence in cases of disagreement over contractual interpretations.

For example, systematically recording decisions taken in steering meetings, adjustments to action plans, and formal exchanges between parties makes it easier to trace decision histories and pinpoint divergence points. Such archives, accessible at any time, are valuable tools for legal experts and mediators involved in potential disputes.

5.3.4 Conflict-Management Training

Finally, to reduce the occurrence of disputes and promote a stable working environment, teams should be regularly trained in conflict management. Dedicated training programs help participants recognize early warning signs, adopt de-escalation strategies, and develop assertive communication skills.

These programs may take the form of interactive workshops, conflict-simulation exercises, or one-on-one coaching sessions. By improving understanding of interpersonal dynamics, stakeholders are better equipped to anticipate and resolve tensions before they undermine collaboration quality.

5.4 Stakeholder Awareness and Training

The long-term performance of hydrocarbon assets largely depends on the competence and commitment of stakeholders. Raising awareness and providing continuous training—

whether for technical teams, government representatives, or local communities—are therefore essential levers to ensure responsible and sustainable exploitation.

5.4.1 Awareness of Environmental and Social Issues

Hydrocarbon activities have significant impacts on the environment and local communities. It is thus imperative to raise awareness among all stakeholders of the associated environmental and social challenges.

Awareness campaigns, supported by targeted communication programs, help highlight the importance of adopting responsible practices. For example, seminars on waste management, emissions reduction, or biodiversity protection can be organized jointly by operators and public authorities. Such initiatives strengthen the social legitimacy of projects and build trust with affected communities.

5.4.2 Technical Training Programs

Developing local technical skills is a major challenge for ensuring autonomy and sustainability of hydrocarbon projects. Technical training programs not only enhance the expertise of local teams but also promote technology transfer and dissemination of best practices.

These programs may take various forms: practical workshops on using new technologies, specialized seminars on preventive equipment maintenance, or training on industrial risk management. By involving international experts and collaborating with academic institutions, asset owners and operators can help strengthen local capabilities, ensuring more efficient and safer resource management.

5.4.3 Strengthening Institutional Capacities

Government agencies and institutions overseeing operations must have the necessary expertise to enforce regulations and ensure compliance with standards. Strengthening institutional capacity involves continuous staff training, modernizing monitoring tools, and improving control mechanisms.

For example, establishing a training center dedicated to energy and environmental issues can provide regulators with the knowledge and tools to conduct in-depth evaluations of hydrocarbon projects. This not only improves monitoring quality but also fosters constructive dialogue with operators, grounded in mutual understanding of constraints and requirements.

5.4.4 Involvement of Local Communities

Finally, involving local communities in hydrocarbon projects is essential for socially responsible and harmonious resource management. By integrating local populations into decision-making processes and offering them a voice on projects, authorities strengthen social cohesion and facilitate greater acceptance of operations.

Community forums or regular public consultations provide opportunities for residents to express concerns, suggest improvements, and participate actively in implementing environmental-protection measures. This participatory approach builds trust and establishes a lasting partnership between operators, governments, and local communities. It also helps maximize economic and social benefits by promoting local employment, community infrastructure development, and implementation of social-responsibility programs.

Key Takeaways

Chapter 5 highlights the crucial importance of collaboration between asset owners—often States or public entities—and private operators responsible for hydrocarbon exploitation. It emphasizes that project success depends on building mutual trust and transparent communication, reinforced by regular meetings and use of collaborative technological tools. It stresses the need for shared commitments, notably through negotiation frameworks that integrate national interests, conflict-resolution mechanisms, and continuous training processes to strengthen stakeholder competencies. Through this integrated approach, the chapter demonstrates how effective cooperation—supported by rigorous procedures and precise documentation—maximizes the value of petroleum assets while ensuring sustainable exploitation in a complex and evolving environment.

Chapter 6

Strategies for Continuous Improvement in the Management of NOPAs

The management of NOPAs occupies a strategic place in the hydrocarbons sector. These assets, though held without being directly operated, require a specific managerial approach that combines operational rigor, technological innovation, and proactive risk management. Continuous improvement therefore represents a fundamental lever to ensure not only the competitiveness and profitability of such investments but also their sustainability in a constantly evolving environment. This chapter explores in depth the key strategies for embedding a culture of continuous improvement in the management of NOPAs. It examines methods for implementing integrated management systems, developing skills and continuous training, as well as mechanisms for monitoring and evaluating performance. It also addresses the sector's inherent challenges and emerging opportunities, highlighting the importance of benchmarking, adoption of technological innovations, and the strengthening of administrative frameworks.

6.1 Implementation of Integrated Management Systems

6.1.1 The Need for a Holistic Approach

In a sector as complex as hydrocarbons, NOPAs involve multiple stakeholders—operators, investors, and regulatory authorities. To optimize coordination and communication among them, the establishment of an integrated management system is crucial. Such a system must centralize information, standardize procedures, and ensure full transparency across all processes. This holistic vision promotes faster, better-informed decision-making while reducing risks of errors and redundancies.

6.1.2 Data and Information Management

Data lies at the heart of any continuous-improvement strategy. Collecting, processing, and analyzing data makes it possible to measure asset performance, detect malfunctions, and identify optimization opportunities.

- **Creation of centralized platforms**

The first step is to establish a centralized platform consolidating all data relating to operational performance, costs, KPIs, and environmental parameters. Such an integrated solution breaks down information silos, harmonizes data sources, and facilitates benchmarking across assets. For example, such a platform may collect real-time data on production, per-well operating costs, or levels of compliance with environmental standards.

- **Development of dynamic dashboards**

Interactive, customizable dashboards are indispensable tools for instant performance visualization. They can display real-time indicators such as monthly production trends, operational cost fluctuations, or equipment failure rates. With such visualizations, managers can quickly identify deviations from objectives and implement corrective measures. Moreover, automated alerts can flag any anomaly detected in the system.

6.1.3 Integration of Operational and Administrative Processes

The effectiveness of an integrated management system depends on harmonizing its constituent processes—both field operations and administrative monitoring and control.

- **Standardization of procedures**

Standardizing practices—whether in performance evaluation or expenditure management—is essential to build a culture of excellence. For example, adopting a uniform process for approving investment expenditures ensures better traceability of financial flows and strengthens budgetary control. Likewise, standardized procedures for risk analysis improve incident anticipation and enable swift implementation of preventive measures.

- **Optimization of communication**

Smooth communication among stakeholders—operators, investors, regulators—is a key success factor. Collaborative platforms and regular meetings align expectations, share critical information, and resolve emerging issues quickly. This collaborative approach fosters responsiveness and decision-making adapted to sector challenges.

6.1.4 Risk Management and Compliance

By their very nature, NOPAs carry particular risks—financial, operational, and environmental. Robust risk-management mechanisms are therefore indispensable to preserve value and sustainability.

- **Risk-management mechanisms**

The first step is to identify, assess, and mitigate risks using specialized tools. For example, scenario analyses help anticipate the potential impacts of an oil-price drop or environmental incident. Such analyses provide forward-looking insights that support contingency planning and mitigation strategies. In addition, simulation and modeling software can test different hypotheses and optimize responses to unforeseen events.

- **Regulatory compliance**

Compliance with local and international standards is a critical issue for NOPAs. Non-compliance can lead to heavy financial penalties and reputational damage. To ensure ongoing compliance, internal policies must be regularly updated to reflect legislative changes. Regular internal and external audits complete this framework, providing continuous oversight and quickly detecting any non-compliance.

6.2 Skills Development and Continuous Training

Human resources are at the core of performance in managing NOPAs. In a constantly evolving sector, skills development and continuous training of teams are indispensable for maintaining agile and effective management.

6.2.1 Stakeholder Training

Team training must be approached from two complementary angles: technical and strategic.

- **Technical aspects**

Technological advances in hydrocarbons demand constant updating of technical skills. Operators and managers must be trained in the use of the latest tools—geological modeling software, real-time monitoring systems, or advanced data-analysis platforms. Such technical training not only improves operational efficiency but also reduces human error and the risk of incidents.

- **Strategic aspects**

Beyond technical skills, it is essential to develop capabilities in strategic management, particularly in contract negotiation, stakeholder relations, and regulatory compliance. Specific training in these areas enables managers to navigate complex environments and adopt strategies suited to sector challenges. For instance, negotiation-simulation workshops can prepare teams to handle crisis situations or delicate negotiations with external partners.

6.2.2 Strengthening Interdisciplinary Collaboration

Managing NOPAs requires a broad set of skills drawn from engineering, finance, risk management, and environmental science. Encouraging interdisciplinary collaboration is therefore crucial to achieving a comprehensive view of issues and optimizing decision-making.

- **Cross-disciplinary exchanges**

Establishing interdisciplinary working groups encourages knowledge sharing and diverse perspectives. For example, regular meetings bringing together engineers, financiers, and environmental specialists can reveal unexpected synergies and adapt strategies based on collective experience.

- **Mentorship programs**

Implementing mentorship programs supports knowledge and skill transfer across generations of experts. Experienced managers can guide young professionals in learning industry best practices while offering strategic insights grounded in real-world experience.

6.3 Continuous Improvement through Monitoring and Evaluation

A rigorous monitoring and evaluation framework forms the backbone of any continuous-improvement initiative. Such a system makes it possible to measure the performance of NOPAs, identify weaknesses, and implement effective corrective actions.

6.3.1 Key Performance Indicators (KPIs)

Defining and tracking KPIs are indispensable tools for measuring the effectiveness of implemented actions.

- **Operational performance**

Among operational KPIs are indicators such as production rates, equipment efficiency, and operating costs per barrel. These metrics enable real-time assessment of facility performance and early detection of anomalies before they escalate into major problems. For example, a sudden drop in production may indicate a technical malfunction requiring immediate intervention.

- **Financial performance**

Financial indicators such as return on investment, operating margins, and unexpected expenses provide a clear picture of the economic health of assets. Continuous monitoring of these indicators enables adjustments to investment strategies and ensures sustainable profitability. In a volatile environment, the ability to control costs and optimize expenditures represents a critical competitive advantage.

- **Environmental performance**

Environmental concerns are central to current priorities. Monitoring water consumption, greenhouse gas emissions, and waste management helps assess the ecological impact of operations. These environmental KPIs are vital not only for regulatory compliance but also for meeting stakeholder expectations on sustainable development.

6.3.2 Regular Audits and Lessons Learned

Audits are valuable tools for assessing process effectiveness and verifying compliance with established standards.

- **Internal and external audits**

Conducting regular internal and external audits provides an in-depth diagnosis of management practices. These audits offer an objective view of organizational strengths and weaknesses, allowing quick identification of improvement areas. Moreover, external audits enhance the company's credibility with investors and regulators.

- **Lessons learned**

Analyzing lessons from past incidents or successes is an essential source of organizational learning. Documenting both positive and negative experiences enables the organization to build on these lessons, adjust strategies, and avoid repeating mistakes. Regular review meetings, supported by detailed reports, help disseminate lessons across teams and foster a culture of continuous improvement.

6.4 Overcoming Challenges and Seizing Opportunities

Managing NOPAs is not without challenges. Between dependency on operators and limited control over daily operations, managers face complex situations that demand flexibility and foresight. However, these challenges also bring major strategic opportunities.

6.4.1 Inherent Challenges in Managing NOPAs

- **Limited control over operations**

Unlike in-house-operated assets, NOPAs rely heavily on the expertise and management of third-party operators. This model may lead to a lack of direct control over day-to-day operations, complicating rapid corrective action in case of malfunctions. Strengthening monitoring mechanisms and communication with operators is therefore essential to ensure compliance with quality and safety standards.

- **Contractual dependency and associated risks**

Contractual relationships with operators can create tensions, especially in cases of diverging interests or disagreements over exploitation strategy. Managing contractual relationships thus becomes a major challenge, requiring careful contract drafting and the establishment of mediation mechanisms to resolve potential conflicts.

6.4.2 Strategic Opportunities

- **Optimized management of operator relations**

Despite the challenges, proactive management of contractual relationships can transform obstacles into opportunities for collaboration. Clear, detailed agreements, supported by performance clauses, strengthen trust between parties. Regular, transparent communication with operators fosters joint solutions to difficulties, turning the relationship into a true strategic partnership.

- **Leveraging synergies among investors**

NOPAs create opportunities to build strategic partnerships with other investors or operators. Such collaborations provide access to new technologies, allow risk-sharing, and generate economies of scale. For example, a consortium of investors can pool infrastructure-investment costs while benefiting from the diverse expertise of partners. This collaborative approach leads to more efficient management and reduced operating costs.

- **Diversification of asset portfolios**

Managing NOPAs also facilitates diversification of investments. By incorporating assets across different geographic zones or under varied regulatory regimes, managers can mitigate risks associated with market volatility or legislative changes. This diversification enhances portfolio resilience and ensures long-term investment sustainability.

6.5 Benchmarking and International Best Practices

Adopting best practices from international experience is a key driver of continuous improvement in managing NOPAs. Benchmarking—comparing organizational performance with that of industry leaders—helps identify gaps and adopt proven solutions.

6.5.1 Identifying Industry Leaders

- **Comparative performance analysis**

It is essential to identify companies excelling in NOPAs management and study their strategic approaches. This process highlights best practices in cost control, performance monitoring, and risk management. For example, detailed analysis of annual reports and published case studies from these leaders can provide concrete improvement pathways tailored to each organization's context.

- **Participation in international conferences**

International conferences and seminars offer prime opportunities to exchange with industry experts. Participation in such events provides exposure to innovative ideas, emerging technologies, and collaborative networks. These exchanges enhance managers' ability to anticipate market developments and adapt strategies accordingly.

6.5.2 Adoption of Best Practices

- **Creation of a best-practices repository**

Developing a guide that consolidates best practices in NOPAs management provides a reference tool for all teams. This repository can include standardized procedures for performance evaluation, cost-reduction strategies, and risk-analysis methodologies. Regular distribution of this guide fosters a culture of excellence and harmonizes practices within the organization.

- **Conducting comparative audits**

Comparative audits measure the gap between organizational performance and that of industry leaders. These audits provide an objective assessment of internal processes and highlight improvement opportunities. By identifying areas where efficiency gains are possible, companies can adopt proven solutions to optimize management.

6.6 Continuous Training for Administrative Executives

Beyond the technical training of operational teams, strengthening the skills of administrative executives is also indispensable for ensuring effective management adapted to sectoral evolutions.

6.6.1 Technical and Managerial Training

- **Updating technical skills**

In a constantly evolving technological environment, administrative executives must remain at the forefront of new technologies. Regular training on management software, advanced data analysis, and risk modeling improves responsiveness and decision-making accuracy. Online training modules provide greater flexibility, allowing executives to develop skills while maintaining their daily responsibilities.

- **Developing managerial skills**

The ability to manage teams effectively, negotiate with external partners, and lead complex projects is essential in managing NOPAs. Training in management, conflict resolution, and leadership strengthens these skills and equips executives with a comprehensive strategic approach. Workshops and practical case studies provide opportunities to apply learned concepts in real-world contexts.

6.6.2 Certifications and Exchange Programs

- **Obtaining recognized certifications**

Encouraging executives to pursue professional certifications such as **PMP (Project Management Professional)** or certifications in financial auditing is a long-term investment. These attest to executive competence and enhance the organization's credibility with investors and strategic partners. Internal incentives, such as training bonuses or career advancement opportunities, can motivate staff to obtain such certifications.

- **Exchange programs and international collaboration**

Collaboration with universities and international organizations helps organize exchange programs and placements, offering executives the opportunity to experience global management practices. Such study missions facilitate the integration of new ideas and strengthen the organization's ability to adapt to international market dynamics. Participation in international seminars and conferences also broadens professional networks and provides fresh perspectives.

6.7 Adoption of Technological Innovations

Digital transformation is increasingly taking hold across industries, and hydrocarbons are no exception. Adoption of innovative technologies is a major lever for optimizing the management of NOPAs and meeting rising performance and sustainability expectations.

6.7.1 Digitization and Automation

- **Integration of advanced digital systems**

Digitizing processes for data collection, analysis, and reporting enhances efficiency and accuracy. Integrating tools such as **ERP systems** and **Business Intelligence platforms** enables real-time monitoring of operational and financial performance. Furthermore, automating administrative and operational tasks reduces human error and eases team workloads.

- **Development of mobile applications and cloud solutions**

Mobile apps for field-operation monitoring provide flexibility and enable instant data collection. Meanwhile, cloud solutions guarantee continuous information accessibility, supporting rapid, collaborative decision-making among stakeholders.

6.7.2 Advanced Data Analytics

- **Implementation of predictive models**

Predictive analysis using forecasting algorithms is a major innovation in NOPAS management. These models anticipate production fluctuations, predict equipment failures, and simulate the impact of different economic scenarios. With such tools, managers can adjust investments and optimize facility maintenance.

- **Exploitation of Big Data**

Processing and analyzing large volumes of data from sensors, monitoring systems, and audit reports provide detailed, real-time views of operations. Implementing Big Data solutions helps uncover hidden trends, optimize resource allocation, and improve responsiveness to unexpected events.

6.7.3 Green and Sustainable Technologies

- **Promotion of green technologies**

In the context of energy transition, integrating environmentally friendly technologies into asset management is essential. Investing in CO_2-reduction solutions, optimizing energy consumption, and managing resources sustainably enhance the environmental performance of operations. Examples include installing heat-recovery systems or using sensors to optimize energy usage at facilities.

- **Investment in renewable energy**

Diversifying energy sources within the portfolio of NOPAs mitigates risks linked to oil-market volatility. Incorporating renewable projects—such as wind or solar—not only provides long-term economic benefits but also reinforces organizational commitment to sustainable development.

6.8 Synthesis and Future Outlook

The management of NOPAs in the hydrocarbons sector sits at a crossroads where operational rigor meets technological innovation and the need for collaborative approaches. Continuous improvement, as a managerial philosophy, proves to be an indispensable tool for optimizing performance, strengthening competitiveness, and ensuring investment sustainability.

6.8.1 A Culture of Continuous Improvement

Embedding a culture of continuous improvement requires creating an organizational framework in which every actor, at all levels, is engaged in the constant pursuit of excellence. Data centralization, process standardization, continuous training, and adoption of innovative technologies form the pillars that, together, transform potential challenges into strategic opportunities.

6.8.2 Evolving Regulatory Frameworks and Market Expectations

The hydrocarbons sector is undergoing profound change, both technologically and in environmental regulation. Evolving standards and increasing stakeholder demands require managers to constantly adapt strategies. Regular audits, policy reviews, and adoption of new technologies ensure NOPAs remain compliant with international standards and competitive on global markets.

6.8.3 Prospects for Innovation and Integration

Technological innovations—particularly in artificial intelligence, Big Data analytics, and process digitization—open unprecedented perspectives for managing NOPAs. In the future, integrating advanced technologies will further optimize operations, reduce costs, and strengthen strategic decision-making. In addition, expanding interdisciplinary collaboration and fostering international partnerships will create a fertile environment for new ideas and adoption of global best practices.

Key Takeaways

Chapter 6 presents an integrated approach to embedding continuous improvement in NOPAs management. It underscores the importance of implementing integrated management systems that centralize information, standardize processes, and optimize stakeholder communication. Emphasis is placed on developing both technical and strategic skills through continuous training and enhanced interdisciplinary collaboration, fostering effective knowledge transfer. The chapter also demonstrates how regular monitoring—via KPIs and internal and external audits—identifies variances and supports adjustments that improve operational and financial efficiency. Finally, it highlights the adoption of technological innovations—such as digitization, automation, and predictive analytics—that turn challenges into strategic opportunities, ensuring competitiveness and sustainability of investments in an ever-evolving environment.

Chapter 7

Basic Practical Rules for Managing an Oil Asset

Managing an oil asset is a strategic issue in the hydrocarbons industry. Whether operated or non-operated, its profitability and sustainability depend on a clear understanding of both economic and operational dynamics. For a manager—especially in the case of a NOPAs—the primary responsibility is to ensure that the asset generates profit and contributes to value creation.

To achieve this, it is essential to rigorously analyze financial flows, closely monitor costs and revenues, and implement short-, medium-, and long-term strategies. This chapter outlines the basic principles of oil-asset management, covering profitability analysis, specific management of mature assets, and the development of reservoir-management plans, while integrating operational-optimization practices and risk management.

7.1 Economic Understanding and Profitability

The first step in managing an oil asset is to establish a clear vision of its profitability. It is not enough to note that the asset produces hydrocarbons; it is equally necessary to assess whether this production translates into net profit after deducting all operating, investment, and maintenance costs.

To do so, several key financial indicators must be monitored:

- **Production cost per barrel**: This parameter evaluates operational efficiency. If costs are too high, the asset may become uncompetitive—even during periods of high oil prices.

• **Selling price and net margins**: Monitoring revenues from hydrocarbon sales and analyzing profit margins ensure that operations remain profitable.

• **Cash flows**: Prudent cash management is essential to anticipate and finance future investments or maintenance operations.

Key questions for monitoring profitability

The manager of a petroleum asset—and particularly of a non operated one—must be able to answer fundamental questions such as:

1. **Is the asset profitable?**

Analysis is required to determine whether revenues exceed both direct and indirect costs associated with production.

2. **Do current investments and expenses create value?**

Cost-benefit analyses help determine whether each expense—whether in maintenance, technology, or safety—translates into short- or long-term economic gains.

Mature assets often present specific challenges. In such cases, profitability does not depend on dramatic production increases, but rather on the ability to control costs and maintain stable operations despite natural production decline. The manager must therefore rigorously track economic indicators and adjust strategies according to both market conditions and the state of the asset.

7.2 Management of Mature Assets

Mature assets—whether late-life oil fields or aging infrastructure—require a specific management approach. While still potentially lucrative if properly handled, they present unique characteristics that demand tailored strategies to extend their lifespan and maximize extracted value.

7.2.1 Definition of a Mature Asset

A mature asset is typically characterized by:

- **Declining production**: Oil-field production tends to plateau or decline over time as natural reserves are depleted.
- **Technological obsolescence**: Equipment and infrastructure may be outdated compared to current standards, sometimes requiring costly upgrades.
- **Rising maintenance costs**: Aging equipment is more complex and costly to maintain, with higher risks of failure.
- **Increased HSE risks**: Mature assets carry heightened risks of failure or environmental incidents, requiring reinforced health, safety, and environmental (HSE) monitoring.
-

7.2.2 Specific Characteristics and Operational Issues

Managing a mature asset relies on a thorough understanding of its particularities and the adoption of operational strategies designed to maximize value while minimizing risks. Key features include:

1. **Declining returns**

Over time, profitability decreases as production falls, while fixed costs remain constant or increase. Continuous operational optimization is necessary to maintain viable economic balance.

2. **Complexity of interventions**

Maintenance and refurbishment work becomes increasingly technical. Interventions on aging infrastructure require specialized engineering expertise and advanced technologies to avoid overly costly or ineffective repairs.

3. **Need for additional investments**

To extend a mature asset's life or enhance performance, significant investments are often required. These must be evaluated through rigorous cost-benefit analysis to ensure measurable gains in production or risk reduction.

4. **Cost-benefit analysis**

Any intervention—whether maintenance work or investment in new technologies—must be justified by tangible improvements in asset performance. The goal is to balance costs, benefits, and risks.

7.2.3 Challenges and Strategies for Managing Mature Assets

The management of mature assets poses several challenges that managers must anticipate and address with appropriate strategies:

7.2.3.1 Maintenance and Renovation

Mature assets require constant monitoring and regular work to remain operational. Maintenance strategies should aim to prevent failures and optimize schedules to reduce downtime. Examples include:

- **Predictive maintenance**: Using sensors and data-analysis technologies to detect early signs of failure and anticipate necessary interventions. This proactive approach reduces maintenance costs and minimizes major incident risks.

- **Optimized maintenance scheduling**: Planning interventions during periods of minimal production impact helps contain costs and ensure continuous operation.

7.2.3.2 HSE Risk Management

Health, safety, and environmental (HSE) considerations become increasingly critical in mature assets. Effective risk management includes:

- **Regular audits**: Frequent inspections verify compliance with standards and identify potential failures early.
- **Investments in regulatory compliance**: Investing in technologies and procedures that ensure HSE compliance helps avoid sanctions, shutdowns, or accidents.
- **Continuous staff training**: Regularly updating staff skills and training on new technologies and safety protocols reduces risks of accidents and environmental incidents.

7.2.3.3 Well-Intervention Optimization

In mature fields, wells often require specific interventions—such as reperforation or stimulation—to improve production. These must be justified through detailed analysis:

- **Assessment of remaining reserves**: Evaluating residual production potential determines whether investment will deliver significant output increases.
- **Cost-benefit study**: Comparing intervention costs with potential production gains ensures economic viability.

7.2.3.4 Cost Reduction and Resource Pooling

To manage rising maintenance costs, several strategies can be implemented:

- **Resource sharing across licenses**: Pooling equipment and intervention teams reduces unit costs and optimizes management across multiple assets.

- **Digitization and automation**: Integrating digital and automated solutions into monitoring and maintenance processes reduces reliance on labor and improves efficiency.

7.2.3.5 Extending Asset Life

Maximizing value from a mature asset often requires investing in technologies that extend its lifespan, such as:

- **Gas or water injection**: These methods maintain reservoir pressure, delaying production decline and extending profitable operations.

- **Use of advanced materials**: Replacing aging components with next-generation materials reduces maintenance costs and improves infrastructure reliability.

7.2.3.6 Best Practices for Sustainable Operational Management

To successfully manage mature assets, a long-term vision and close stakeholder collaboration are essential. Best practices include:

- **Development of a long-term management plan**: Such a plan must account for technological developments, market constraints, and economic goals, supported by rigorous budgeting and prioritization of high-value projects.

- **Stakeholder collaboration**: Cooperation among operators, regulators, investors, and other actors ensures investments and operational strategies align with economic and environmental requirements.
- **Investment in training and skills development**: Upskilling teams on new technologies and HSE standards is critical for ensuring high-quality interventions and asset sustainability.

7.3 Importance of a Reservoir Management Plan

Reservoir management is a central discipline in the hydrocarbons industry. Its goal is to optimize resource development while ensuring reservoir stability and integrity over the long term. A well-designed reservoir management plan is the cornerstone of balanced and sustainable exploitation. Such a plan must integrate technical, economic, and environmental aspects in order to maximize hydrocarbon recovery while minimizing negative impacts on the reservoir.

7.3.1 Objectives of a Reservoir Management Plan

The primary objectives of a reservoir management plan are multiple and interdependent:

- **Maximizing recovery:** The fundamental goal is to optimize hydrocarbon extraction volumes using advanced techniques and appropriate development strategies.
- **Preserving reservoir integrity:** It is critical to avoid any operation likely to damage the reservoir, whether through abrupt pressure variations or inappropriate interventions.
- **Minimizing operating costs:** Effective reservoir management should reduce expenditures by strategically planning interventions and using optimization technologies.

- **Regulatory compliance:** The plan must fully comply with government and regulatory requirements, particularly in HSE (health, safety, and environment).

7.3.2 Pressure Differentials and Reservoir Protection

In many countries, authorities impose strict pressure-differential limits to regulate hydrocarbon extraction. These thresholds are set to protect the reservoir against irreversible damage that could compromise production capacity. Non-compliance can lead to serious consequences:

- **Damage to reservoir rock:** Excessive pressure can fracture the rock, reducing its ability to contain hydrocarbons.
- **Water intrusion:** Excessive pressure decline can trigger water influx into the reservoir, diluting production quality and lowering overall recovery.
- **Equipment overstress:** Abnormal pressure fluctuations can damage surface and subsea infrastructure, leading to unexpected costs and production shutdowns.

Rigorous monitoring and compliance with pressure differentials are therefore essential to preserve reservoir integrity and ensure sustainable exploitation.

7.3.3 Long-Term Planning and Development Scenarios

Sustainable reservoir exploitation requires long-term planning that anticipates technical, economic, and environmental developments. Key elements include:

- **Reservoir modeling:** Advanced software simulates reservoir behavior under different intervention scenarios. These models provide accurate forecasts to guide strategic decisions.

- **Development scenarios:** Preparing alternative development paths allows anticipation of outcomes, whether gradual production decline or the need for specific interventions.
- **Long-term budgeting:** Defining staged budgets to cover future investments and operations ensures the availability of financial resources to implement strategies.
- **Contingency plans:** Backup strategies must be prepared for unforeseen events such as technical incidents or market shocks, ensuring rapid and effective responses.

7.3.4 Continuous Surveillance and Technological Tools

Continuous reservoir monitoring is a pillar of proactive and reactive management. It relies on advanced technological tools that measure critical reservoir parameters in real time:

- **Pressure and temperature sensors:** Installed in wells, these devices continuously measure variations, providing essential data for adjusting operating strategies.
- **Production data analysis:** Regular comparison of actual production rates against forecasts enables early detection of anomalies, facilitating timely decision-making.
- **4D seismic:** Time-lapse seismic (3D plus time) provides a dynamic view of reservoir evolution, highlighting weaknesses and identifying intervention opportunities.

The benefits of continuous monitoring are significant: faster decision-making, failure prevention, and optimized recovery efficiency.

7.3.5 Enhanced Oil Recovery (EOR) Techniques

As reservoirs mature, conventional extraction methods become less effective. Enhanced Oil Recovery (EOR) plays a crucial role in optimizing production. Main EOR techniques include:

- **Gas injection:** Injecting CO_2 or natural gas maintains reservoir pressure and mobilizes residual hydrocarbons.
- **Water injection:** Used to raise reservoir pressure, driving hydrocarbons toward producing wells.
- **Chemical injection:** Adding polymers or surfactants reduces interfacial tension between water and oil, improving hydrocarbon mobilization.

These techniques should be incorporated into the reservoir management plan, with careful assessment of required investments and potential production gains.

7.3.6 Risk Management and Environmental Impact

Reservoir management is not limited to production optimization; it also requires robust management of technical, economic, and environmental risks. Recommended approaches include:

- **Regular audits:** Periodic inspections ensure infrastructure compliance and enable early anomaly detection.
- **Continuous training:** Skill enhancement through adapted training programs ensures teams can respond quickly and effectively to incidents.
- **Interdisciplinary collaboration:** Combining expertise from geologists, engineers, economists, and HSE specialists fosters a holistic approach to reservoir management.

Environmental impacts must also be constantly evaluated and controlled:

- **Emission reduction:** CO_2 capture and storage technologies can help limit the carbon footprint of operations.
- **Waste management:** Proper treatment of fluids and waste is vital to prevent soil and water pollution.
- **Site restoration:** After production ends, remediation plans should be implemented to rehabilitate affected areas.

Synthesis and Perspectives

Managing an oil asset—whether non operated or mature—requires an integrated, multidimensional approach. It depends on the ability to reconcile economic, operational, and environmental imperatives in contexts often marked by uncertainty and technical complexity.

Key Points to Remember

1. **Economic understanding and profitability tracking**

Rigorous analysis of production costs, margins, and financial flows is the first step to ensuring the asset creates value. Proactive management detects early signs of declining profitability and allows timely strategy adjustments.

2. **Specific management of mature assets**

Mature assets require tailored strategies to address production decline and aging infrastructure. Predictive maintenance, optimized scheduling, and resource pooling are key levers for extending asset life and maintaining economic performance.

3. **Developing a reservoir management plan**

A robust plan integrates long-term planning, continuous monitoring of critical parameters, EOR techniques, and rigorous risk management. It forms the basis of balanced exploitation—maximizing recovery while preserving reservoir integrity.

4. **Integrating HSE and sustainability**

Safety, health, and environmental stewardship are fundamental dimensions of oil-asset management. Regulatory compliance, continuous training, and eco-friendly technologies minimize incident risks and ensure sustainable operations.

Future Perspectives

The evolution of digital technologies, improvements in monitoring tools, and the integration of cutting-edge technical solutions open new horizons for oil-asset management. These innovations will not only optimize recovery and reduce costs but also strengthen consideration of environmental and safety concerns. The future of oil-asset management will therefore rest on an increasingly integrated approach—combining technical expertise, economic analysis, and commitment to sustainable development.

Managing an oil asset, whether in its initial exploitation phase or in maturity, requires a deep understanding of economic and operational dynamics. Monitoring financial indicators, implementing predictive maintenance strategies, rigorous HSE risk management, and developing a reservoir management plan are all elements that, when consistently applied, ensure asset profitability and sustainability.

The above guidance provides a comprehensive reference framework for oil-asset managers. It underscores the importance of an integrated approach—combining technical analysis, economic oversight, and compliance with environmental requirements. In the context of energy transition and market volatility, continuous adaptation of management strategies is indispensable to maintain competitiveness and operational safety.

Thus, whether through technological investments, process digitization, or continuous staff training, every action must aim to optimize the overall performance of the asset. Successful asset management rests on balancing hydrocarbon recovery maximization with risk minimization, while guaranteeing environmentally responsible operations.

At a time when energy and environmental challenges are becoming increasingly urgent, managers' ability to anticipate and innovate in their practices will determine the success of oil projects. The way forward requires synergy between advanced technologies, operational expertise, and a commitment to sustainable exploitation—transforming the challenges of mature assets into long-term value-creation opportunities.

Consequently, understanding economic dynamics, mastering technical interventions, and embedding rigorous management of environmental and safety risks constitute the fundamental pillars of effective oil-asset management. By presenting recommended practices and optimization strategies in detail, this chapter provides managers with the tools needed to address both current and future challenges in a constantly evolving sector.

Ultimately, effective oil-asset management relies on an integrated approach, long-term vision, and adaptability to technological and regulatory developments. By combining financial rigor, technical expertise, and environmental commitment, operators and owners can guarantee profitable and sustainable exploitation aligned with contemporary expectations of competitiveness and responsibility.

Key Points to Remember

Chapter 7 outlines the fundamental practical rules for effective oil-asset management. It details how to rigorously analyze asset profitability by examining production costs, profit margins, and cash flows, ensuring that hydrocarbon exploitation creates real economic value. It also addresses the specific management of mature assets, highlighting challenges related to production decline and rising maintenance costs, and stresses the need to adapt intervention strategies to extend asset life. Furthermore, it underscores the necessity of developing a reservoir management plan that optimizes recovery, preserves reservoir integrity, and integrates enhanced oil recovery techniques. Long-term planning, cost-benefit analysis, and continuous monitoring of technical and economic parameters are presented as essential levers for ensuring sustainable and secure exploitation of petroleum resources.

Chapter 8

Contractual Dynamics and Rent Sharing

The exploitation of petroleum resources rests on complex legal agreements signed between governments—or their state-owned entities—and oil companies, often structured as consortiums. These contracts, which serve as true instruments of regulation, are designed to define in detail the modalities of risk, cost, and benefit sharing related to hydrocarbon exploration, development, and production within designated geographical areas. They reflect, on the one hand, the economic objectives of States—eager to maximize financial returns and guarantee sovereignty over their resources—and, on the other hand, the imperatives of profitability, legal security, and return on investment demanded by private investors. In a context of globalization and intensified competition, the evolution and sophistication of such contracts play a decisive role in optimizing rent distribution and ensuring the long-term sustainability of oil assets.

8.1 Types of Petroleum Contracts

Petroleum contracts take several forms, with the choice largely depending on subsurface characteristics, economic stakes, and the strategic priorities of States and operators. The main contractual models include:

8.1.1 Concessions

In concession agreements, the oil company is granted exclusive rights to explore, develop, and produce hydrocarbons within a clearly defined geographical area. This model gives the operator broad freedom in asset management, while obliging it to bear all risks and operational costs.

In return, the company receives a significant share of the revenues generated by production, subject to payment of taxes, royalties, and sometimes even profit-sharing in favor of the State. This system stimulates private investment while preserving a sustainable source of revenue for the State.

However, rigorous monitoring and full transparency in cost management are crucial to prevent abusive practices that could undermine the economic balance of the agreement.

8.1.2 Production Sharing Contracts (PSCs)

Production Sharing Contracts are another widely used model in the oil industry. Under this framework, the company initially assumes all exploration and development costs. Once production begins, a portion—called **cost oil**—is allocated to the operator to recover its investment and operating costs. The remainder—known as **profit oil**—is then shared between the company and the State according to pre-negotiated percentages.

This mechanism offers a dual advantage: it reduces the State's financial exposure while incentivizing the operator to optimize costs and improve project profitability. However, striking the right balance between cost recovery and profit sharing remains a major negotiation issue, one that must be adapted to oil-price fluctuations and sustainable development imperatives.

8.1.3 Service Contracts

Service contracts adopt a different approach in that the operator does not necessarily own the hydrocarbons extracted. Here, the company acts as a service provider to the State, supplying technical expertise and operational capacity to carry out exploration and production.

The operator's remuneration is based on a fixed fee or a predetermined share of revenues, while the State retains full ownership of the petroleum produced. This model reduces the State's exposure to operational risks, though cost management and assurance of service quality remain critical issues to guarantee transparency and efficient exploitation.

8.1.4 Joint-Venture Contracts

Joint-venture contracts establish strategic partnerships between a State entity and one or more private companies. This model is characterized by shared investments, risks, and benefits, thereby ensuring a more balanced distribution of responsibilities and better resource management.

Combining the technical and financial expertise of private operators with the local knowledge and negotiating power of the State creates a synergy that promotes more efficient asset exploitation. However, this type of collaboration requires robust governance mechanisms to ensure alignment of interests and prevent potential conflicts over task distribution and benefit sharing.

8.2 Cost Saturation and Budgetary Control

One of the major challenges of contractual management in the oil sector is controlling operational costs. The notion of cost saturation refers to the tendency of certain members—especially non-operating partners—to accept high expenditures in order to maximize short-term revenues. While this approach may seem advantageous in the immediate term, it often proves counterproductive in the long run when such costs fail to generate real value creation or significantly improve the asset.

To counter this drift, contracts must include strict budgetary-control mechanisms. Among these are clauses requiring prior approval of expenditures, regular audits by independent experts, and adoption of cost-validation procedures. These tools not only ensure financial transparency but also encourage rigorous and responsible management—guaranteeing that every investment genuinely contributes to enhancing the value of the oil asset.

8.3 Contractual Performance and Long-Term Objectives

To ensure the sustainable and profitable exploitation of petroleum resources, it is essential to include performance clauses in contracts. These clauses should define clear long-term objectives—ranging from maintaining or increasing production to reducing operating costs and minimizing environmental impact.

Particular attention must be paid to the balance between short-term remuneration and long-term performance goals. Excessive focus on immediate results may encourage operators to adopt unsustainable management strategies, including practices that compromise asset integrity or generate hidden long-term costs.

For this reason, it is advisable to establish Key Performance Indicators (KPIs) to regularly assess operational performance. Such indicators may include daily production rates, average cost per barrel, or hydrocarbon recovery factors. Implementing regular monitoring and evaluation systems supports informed decision-making and allows strategies to be adjusted in line with market changes and technical conditions.

8.4 Impact on Revenues and Conflict Management

The contractual performance of an operator directly affects revenue distribution among partners. Poor management or inefficient exploitation can result in significantly reduced production, unjustified cost increases, or disputes over profit sharing. The consequences go beyond financial aspects, also eroding trust among partners and undermining contract stability.

To mitigate these risks, proactive control and monitoring mechanisms are crucial. For example, periodic audits by independent bodies allow anomalies in cost management or operations to be detected early. In addition, establishing monitoring committees composed of operator and non-operator representatives fosters smooth communication and collective decision-making. By bringing together technical and financial experts, these committees play a key role in quickly resolving conflicts and adjusting strategies to field realities.

Transparency in financial and operational reporting is another critical factor in maintaining partner trust. Regular reports—detailing production trends, cost breakdowns, and operational performance—are indispensable tools for ensuring sound contract management and anticipating potential deviations that could impact revenue distribution.

8.5 Governance Mechanisms and Alignment of Interests

A fundamental challenge in petroleum contracts is aligning the interests of all stakeholders. Achieving this goal is essential to ensure balanced and sustainable exploitation, and it can be pursued through several governance mechanisms:

8.5.1 Fair Sharing of Risks and Benefits

Contract clauses should clearly and transparently define the distribution of risks and benefits. In Production Sharing Contracts, for example, the distinction between cost oil and profit oil must be rigorously established to guarantee that the operator can recover its investments without undermining State revenues. Balanced profit-sharing motivates all parties to act in the best interest of the project, ensuring both cost control and production optimization.

8.5.2 Collaborative Decision-Making Structures

The creation of collaborative governance structures—such as management committees or supervisory boards—ensures the involvement of all partners in the decision-making process. These bodies, composed of State representatives and private companies, foster a collective approach to strategic and operational decisions. By giving every party a voice, they strengthen mutual accountability and reduce tensions arising from divergent interests. Transparency and regularity of committee meetings are key assets for ensuring equitable joint management of the petroleum asset.

8.5.3 Independent Monitoring and Evaluation

To guarantee operational credibility and reliability, the inclusion of independent audits is a recommended practice. Such audits verify cost compliance and contractual execution while providing an objective assessment of operational performance. By relying on third-party experts, stakeholders benefit from impartial oversight, which strengthens transparency and trust among partners.

8.5.4 Inclusion of environmental and social clauses

In today's context of energy transition and heightened environmental concerns, petroleum contracts must integrate clauses addressing environmental protection and social responsibility. These provisions—ranging from greenhouse-gas reduction commitments to ecological-compensation measures—help frame resource exploitation sustainably. They also meet investor expectations, increasingly shaped by ESG (Environmental, Social, and Governance) criteria, and reinforce the legitimacy of petroleum projects within the international community.

8.6 Importance of Audits and Reporting in Contractual Transparency

Transparency is one of the pillars on which successful contractual relationships in the oil sector are built. Rigorous monitoring, coupled with regular audits, ensures operational compliance and prevents potential deviations in cost and revenue management. Key practices in this regard include:

8.6.1 Cost Audits and Financial Control

Every cost incurred in petroleum operations must be fully justified and subject to independent review. Cost audits not only verify the relevance and contractual compliance of expenditures but also help detect anomalies and possible overruns. Establishing strict expense-validation procedures, combined with periodic audits, guarantees transparent financial management and prevents overbilling or abusive cost-optimization practices.

8.6.2 Regular Reporting and Results Communication

The production of regular financial and operational reports is indispensable for maintaining a clear overview of project performance. These reports should detail production trends, cost breakdowns, achievement of performance indicators, and the impact of environmental measures. They are valuable tools for stakeholders, enabling them to adjust strategies based on results and reinforcing trust among partners. Smooth and regular communication also facilitates early anticipation of difficulties and timely implementation of corrective measures.

8.6.3 Monitoring Key Performance Indicators (KPIs)

Defining and tracking KPIs is essential to guide operational project management. Examples include:

- **Daily production:** Measurement of oil volumes extracted over a given period, directly reflecting operational efficiency.
- **Cost per barrel:** A crucial indicator for evaluating economic efficiency and identifying optimization levers.
- **Recovery factor:** A measure of the effectiveness of applied extraction techniques, directly influencing project profitability.

By ensuring regular KPI monitoring, stakeholders have a management tool that allows them to adjust operations according to strategic objectives and anticipate performance gaps.

The evolution of petroleum contracts is part of a continuous process of adaptation to economic, environmental, and technological challenges. Contractual dynamics and rent sharing rest on principles of fairness, transparency, and collaboration among partners. The various contractual models—whether concessions, production sharing agreements, service contracts, or joint ventures—illustrate the diversity of approaches available to optimize petroleum-resource exploitation.

Faced with challenges such as cost saturation, the need for rigorous monitoring, and the integration of long-term performance criteria, it is imperative to establish robust governance mechanisms. Fair sharing of risks and benefits, collaborative decision-making structures, independent audits, and environmental clauses are all levers that help align stakeholders' interests.

Ultimately, transparency in reporting and proactive cost management not only guarantee the project's economic viability but also ensure compliance with environmental and social commitments. In a constantly changing sector, adapting petroleum contracts to new market demands and sustainability imperatives remains a sine qua non condition for ensuring the success and legitimacy of hydrocarbon exploitation.

In conclusion, contractual dynamics and rent sharing are central pillars of petroleum asset governance. They make it possible to reconcile the profitability objectives of private investors with the imperatives of sustainable development and State sovereignty, laying the foundations for a balanced, transparent, and responsible exploitation of natural resources. This constantly evolving contractual framework remains the guarantor of fruitful partnerships and the long-term valorization of petroleum assets within an ever-changing economic and environmental context.

Key Takeaways

Chapter 8 offers an in-depth examination of the structure and evolution of petroleum contracts, highlighting the complexity of the agreements concluded between governments and oil companies for the exploitation of resources. It shows how these contracts—often structured in the form of consortiums—precisely define the allocation of risks, costs, and benefits throughout the phases of exploration, development, and production.

The chapter notably distinguishes between:

- **Concessions**, where the operator assumes all risks in exchange for a significant share of revenues, subject to taxes and royalties; and
- **Production Sharing Contracts (PSCs)**, under which the investor first recovers its costs through *cost oil* before profits are distributed.

The analysis underscores that the continuous refinement and sophistication of these contractual instruments are essential for enabling States to maximize financial returns while ensuring the profitability and sustainability of private investments in a competitive, fast-changing global environment. Thus, contractual dynamics emerge as a major strategic lever for optimizing rent distribution and ensuring the balanced and sustainable management of petroleum assets.

Chapter 9

Core Financial Indicators

In the oil sector—and particularly in the management of NOPAs—financial analysis plays a central role. Core financial indicators provide insight into the economic performance of an asset, guide strategic decision-making, and maximize long-term profitability.

This chapter offers an in-depth study of the main financial tools used to evaluate the viability of NOPAs projects, emphasizing the complementarity of these indicators and their role in risk and investment management.

9.1 Net Present Value (NPV)

Net Present Value (NPV) is undoubtedly one of the most widely used indicators for assessing project profitability. It is defined as the discounted sum of future cash flows generated by the asset, minus initial investments.

The value of NPV lies in its ability to incorporate the time value of money: one dollar today is worth more than one dollar in the future.

In the context of a NOPAs, NPV makes it possible to verify whether projected benefits, once discounted, cover all costs incurred—including the initial investment and operating expenses.

Example: Consider a field expected to generate total revenues of USD 100 million over ten years, with operating costs of USD 60 million and an initial investment of USD 30 million. By applying a discount rate of 10%, the NPV provides a concise picture of the project's profitability. A positive NPV signals value creation; conversely, a negative

NPV suggests that future revenues will not offset costs—indicating the need to revisit the strategy or reassess project assumptions.

9.2 Internal Rate of Return (IRR)

The Internal Rate of Return (IRR) is another key tool for investment evaluation. It corresponds to the discount rate at which NPV equals zero. In other words, IRR represents the average annual return expected from the project, taking into account all cash flows over its lifecycle.

For NOPAs, IRR is particularly useful for comparing different investment opportunities. When a project shows an IRR above the cost of capital, it is generally considered attractive, as it should generate returns exceeding investors' minimum expectations.

This criterion helps managers prioritize investments by directing financial resources toward the most promising projects, while remaining mindful of the inherent risks of oil exploration and production.

9.3 Debt Coverage Ratio (DOC)

The Debt Coverage Ratio (DOC) is an essential indicator for measuring a project's capacity to meet its financial obligations. It is calculated by dividing operating cash flow by debt service (principal repayments plus interest).

A high DOC indicates financial robustness, as it demonstrates that the revenues generated by the asset are sufficient to cover debt-related charges.

For NOPAs, monitoring the DOC is crucial to reassure investors and financial institutions providing credit. Careful debt management mitigates financial risks and ensures economic stability for the project—even in the face of unfavorable oil-price fluctuations.

9.4 Reserve Replacement Ratio (RRR)

The Reserve Replacement Ratio (RRR) measures a company's ability to replace produced reserves with new discoveries or acquisitions. An RRR greater than 1 is generally desirable, as it indicates that the company is replenishing its reserves and ensuring long-term sustainability of its asset portfolio.

For NOPAs, tracking RRR is vital for long-term planning. It confirms whether exploration and development strategies are effective and whether the company is successfully identifying new fields or acquiring promising assets.

An insufficient RRR may prompt a reassessment of investment policy and intensification of exploration efforts to avoid future production decline.

9.5 Compound Annual Growth Rate (CAGR) of Production

The Compound Annual Growth Rate (CAGR) measures the average rate of production growth over a given period. This indicator is particularly relevant for NOPAs, as it provides a clear view of medium- and long-term production trends.

A positive CAGR reflects steady production growth, translating into progressively higher revenues and cash flows.

CAGR analysis allows managers to track the effectiveness of measures implemented to optimize extraction and to quickly identify any decline in performance. In an environment where production may be influenced by technical or geopolitical factors, CAGR serves as a dynamic tool for assessing the operational health of the project.

9.6 Cash Flows and Free Cash Flow (FCF)

Cash flows are the fundamental indicator for measuring the liquidity generated by an asset. They represent inflows and outflows of cash from operations, investments, and financing. Free Cash Flow (FCF) is distinguished in that it corresponds to the cash available after deducting the capital expenditures necessary to maintain or expand the asset.

In the management of NOPAs , analyzing cash flows—and in particular FCF—provides an accurate picture of a company's ability to finance operations and invest in new opportunities without systematically resorting to external debt. A positive and growing FCF is a sign of financial strength, allowing the company to distribute dividends, reduce debt, and strengthen equity.

9.7 Balance Sheet Analysis

The balance sheet is a snapshot of an asset's financial position at a given point in time. It records all assets held, liabilities incurred, and equity. For non operated petroleum assets, regular balance-sheet analysis is indispensable for monitoring debt levels and assessing overall financial health.

Optimizing the balance sheet involves strict cost management, reducing short-term debt, and structuring capital in a balanced way. By identifying performance levers and implementing optimization strategies, managers can strengthen the asset's financial resilience and improve its ability to invest in new projects.

9.8 Return on Revenue (ROR)

Return on Revenue (ROR) measures asset profitability by comparing net profit to revenues generated. This indicator is essential for assessing the efficiency of operational management—especially in NOPAs, where operating and production costs must be carefully controlled to secure sufficient margins.

A high ROR reflects efficient management, where revenues not only cover costs but also generate significant added value for the company. By monitoring ROR regularly, decision-makers can adjust pricing policies, streamline expenditures, and maintain strong competitiveness in a volatile market.

9.9 Return on Capital Employed (ROCE)

Return on Capital Employed (ROCE) quantifies the profitability achieved relative to the total capital committed to the project. This indicator highlights optimal use of financial resources and enables performance comparisons across different assets or projects.

For NOPAs, ROCE carries strategic importance in identifying the most profitable investments. A high ROCE signals rigorous management and efficient capital allocation, while a low ROCE may indicate operational inefficiencies or the need to reassess existing investments. Monitoring ROCE thus guides strategic choices and helps prioritize investments to maximize returns on capital.

9.10 Cost per Barrel Produced

Cost per barrel produced is a crucial operational indicator reflecting the average expenditures required to extract one barrel of oil. This metric is especially relevant in markets characterized by significant price volatility. For NOPAs, controlling this cost is essential to ensuring competitive profit margins.

Reducing cost per barrel involves optimizing extraction processes, adopting innovative technologies, and exercising strict control over operating expenditures. A low cost per barrel enables resilience during global oil-price downturns, thereby enhancing the economic sustainability of NOPAs projects.

9.11 Asset Maturity Index

Finally, the Asset Maturity Index is a synthetic indicator that evaluates the development stage of an asset, from exploration to late-stage declining production. This index positions the asset within its lifecycle and helps orient investment strategies according to its maturity level.

An asset in the exploration or development phase requires significant investment to validate its potential, while an asset in decline may call for different management strategies—such as cost reduction or optimization of residual production. A precise understanding of an asset's maturity thus enables managers to plan next steps, whether by intensifying exploration efforts or preparing a transition to more promising projects.

The management of NOPAs rests on a detailed, multidimensional analysis of financial indicators. From Net Present Value (NPV) and Internal Rate of Return (IRR) to the Debt Coverage Ratio (DOC) and Reserve Replacement Ratio (RRR), each provides unique insight into project performance and economic viability. Integrating these tools into a comprehensive approach enables managers to identify opportunities, prioritize investments, and make informed decisions in a sector where risks are numerous and margins often tight.

Understanding cash flows—both gross and free (FCF)—together with rigorous balance sheet analysis, also contributes to a complete picture of financial health. Likewise, indicators such as Return on Revenue (ROR), Return on Capital Employed (ROCE), and cost per barrel produced provide an operational perspective, indispensable for optimizing performance under fluctuating market conditions.

The Asset Maturity Index complements this framework by situating projects within their lifecycle, allowing investment strategies to be adapted according to the phase of development. The combined use of these indicators offers a holistic approach that improves market foresight, strengthens risk management, and ensures sustainable profitability of non operated petroleum assets.

For investors and managers alike, mastering these financial indicators is not only a technical skill but also a strategic advantage in decision-making. Accurate evaluation of an asset's economic performance allows prioritization of the most promising projects, negotiation of financing on favorable terms, and optimization of resource allocation in an often unpredictable economic environment.

As the oil sector evolves—driven by emerging technologies and the energy transition—the adaptation and updating of these indicators remain essential. They must be embedded in a proactive framework of risk management and financial optimization, ensuring the long-term competitiveness of non operated petroleum assets.

In conclusion, a rigorous and integrated approach to financial indicators transforms the inherent challenges of NOPAs into real investment opportunities. This strategic vision, based on proven tools such as NPV, IRR, DOC, RRR,

CAGR, FCF, balance-sheet analysis, ROR, ROCE, cost per barrel, and the Asset Maturity Index, is an essential lever for navigating a complex and constantly evolving environment. Managers who combine these metrics with contextual analysis and forward-looking vision will be best positioned to maximize profitability and secure the sustainability of their petroleum investments.

Key Takeaway

Chapter 9 underscores the importance of core financial indicators in evaluating the economic performance of NOPAs. It explains how these measures enable investors and managers to determine project profitability, guide strategic decisions, and manage associated risks. Special emphasis is placed on Net Present Value (NPV)—defined as the discounted sum of future cash flows minus the initial investment—which incorporates the concept of the time value of money. The chapter demonstrates that by closely analyzing potential revenues and operating costs, it is possible to assess whether a project can generate long-term value. This financial approach, combined with other indicators, is indispensable for guiding investment in a volatile oil market.

Chapter 10

Integration of Indicators into the Management of NOPAs

The integration of indicators into the management of NOPAs is key to ensuring the sustainability and profitability of investments, particularly in industries as complex and volatile as oil and gas.

By relying on a broad set of financial and operational tools, companies can not only optimize planning and budgeting processes but also strengthen their ability to monitor performance, manage risks, make strategic decisions, and enhance transparency for all stakeholders.

Integrating the various dimensions of these indicators highlights how key metrics—such as Net Present Value (NPV), cash flow, Compound Annual Growth Rate (CAGR), Reserve Replacement Ratio (RRR), Return on Capital Employed (ROCE), Return on Revenue (ROR), the Maturity Index, and other specific ratios (such as DOC or cost per barrel)—contribute to optimized asset management.

Realistic Planning and Budgeting through Financial Indicators

In an environment where market volatility and economic uncertainty have become the norm, planning and budgeting rely on quantitative indicators to ensure project viability.

NPV, in particular, is an essential tool for determining investment profitability by discounting expected future cash flows. Combined with cash-flow analysis, it allows companies to establish realistic budgets by accurately forecasting revenues and expenditures over the project's lifespan.

This approach ensures efficient allocation of financial resources, directing them toward the most promising and strategically relevant projects. It also provides greater visibility on the potential profitability of each initiative, helping minimize the risk of overinvesting in projects with uncertain prospects.

Performance Monitoring: Continuous Tracking of Operational and Financial Indicators

The performance of an asset can only be evaluated by considering a comprehensive set of indicators that measure both operational efficiency and financial profitability.

Monitoring the CAGR is particularly relevant for analyzing medium- and long-term asset evolution, identifying growth trends or periods of slowdown.

At the same time, evaluating ratios such as the RRR, ROCE, and ROR provides complementary insights into financial performance, helping verify whether the asset is generating returns proportional to investments and whether resources are being used optimally.

By combining these measures, managers have a comprehensive dashboard that facilitates performance management and real-time corrective action.

Proactive Risk Management through Detailed Financial Analysis

Market volatility—especially in the oil sector—requires robust tools to anticipate and manage financial risks. Specific ratios, such as the DOC or cost per barrel, quickly flag projects exposed to excessive financial risks.

Through in-depth balance-sheet analyses, managers can detect early warning signs of over-leverage or poor resource allocation. For example, a high cost per barrel may indicate inefficiencies in extraction or processing that could negatively affect overall asset profitability.

Continuous monitoring of risk indicators is therefore essential to prevent financial imbalance and ensure proactive portfolio management.

Strategic Decision-Making Supported by Rigorous Investment Comparisons

In today's globalized, competitive environment, strategic decision-making requires fine-grained analysis of investment opportunities.

Indicators such as the Internal Rate of Return (IRR) and the Maturity Index allow projects to be compared on a consistent basis. IRR provides a precise measure of project profitability by accounting for all discounted future cash flows, while the Maturity Index indicates the stage of development and operational maturity of an asset.

Together, these tools support investment prioritization by identifying projects with the most favorable risk–return profiles. In other words, they help determine which opportunities deserve immediate resource allocation and which can be postponed.

Greater Transparency to Strengthen Stakeholder Trust

Financial transparency is a cornerstone of modern asset management. Clear, accurate communication of financial performance—made possible through systematic integration of indicators—reinforces investor, partner, and stakeholder confidence.

By providing detailed, accessible reports on metrics such as cash flow, NPV, and performance ratios, companies build trust and make it easier to attract additional capital.

Transparency also ensures compliance with rising regulatory demands for corporate governance and financial accountability, while improving coordination among all entities involved in asset management.

Long-Term Optimization in a Volatile Market

Continuous monitoring of financial and operational indicators is crucial to ensuring that assets maintain their value-creation capacity over the long term, even amid volatility.

Oil-price fluctuations, for example, can heavily impact profitability, making regular performance monitoring indispensable. Dynamic analysis of indicators such as ROCE and ROR helps detect potential deviations early and adapt management strategies accordingly.

Furthermore, an integrated approach—combining financial and operational evaluation—provides resilience against market disruptions.

This long-term optimization strategy is based on the principle that continuous process improvement and responsiveness to changes in the economic environment are essential to sustaining value creation.

The Interconnection between Indicators: A Holistic View of Performance

One of the major advantages of integrating indicators into asset management lies in the ability to create synergies between seemingly disparate measures. For example, combining precise cash-flow tracking with NPV analysis not only measures a project's profitability but also forecasts its impact on the company's overall liquidity.

Similarly, correlating the monitoring of return ratios (IRR, ROCE, ROR) with risk-management metrics such as the DOC or cost per barrel provides a holistic view of an asset's financial health. This interconnection between indicators fosters a deeper understanding of both strategic and operational challenges, enabling decision-makers to implement tailored policies and maximize value creation while keeping risks under control.

Implementation Challenges and Future Outlook

Despite the many advantages of integrating indicators into asset management, implementing this approach is not without challenges. One of the main obstacles lies in collecting and analyzing reliable, real-time data. In a sector as complex as oil and gas, it is crucial to have high-performance technological tools capable of handling large volumes of data while ensuring the accuracy of the indicators used.

Moreover, interpreting these indicators requires strong expertise in both finance and operational management, which may be a constraint for some companies. However, technological advances in big data and high-performance software now offer promising perspectives for overcoming these hurdles. These innovations enable automated data collection, improve analytical accuracy, and provide sophisticated visualization tools that facilitate decision-making.

Key Takeaways

Chapter 10 examines the systematic integration of indicators into the management of non operated petroleum assets. It highlights the importance of a combined approach that uses analytical tools and reporting systems to continuously track asset performance, relying on financial and operational indicators such as Net Present Value, return on investment, and various specific KPIs.

This integration allows decision-makers to gain a comprehensive and precise view of project efficiency, supporting risk anticipation and investment-strategy adjustments in line with market fluctuations and environmental constraints.

Overall, the process fosters proactive management aimed at optimizing profitability and ensuring sustainable exploitation in a constantly evolving economic and regulatory environment.

Chapter 11

Economic Evaluation of Petroleum Assets

Purpose of the Chapter

This chapter is somewhat more technical but nevertheless of great importance. It presents the methods commonly used to estimate the economic value of a petroleum asset, particularly in the context of transactions (sales, mergers, equity entry), contractual negotiations, or investment decisions. It applies specially to mature or producing assets.

11.1 Evaluation by Discounted Cash Flows (DCF)

11.1.1 Definition and Principle

The Discounted Cash Flow (DCF) method is a rigorous financial approach used to estimate the Net Present Value (NPV) of a petroleum asset based on expected future net cash flows.

It is based on the principle that the value of an asset today corresponds to the sum of the future cash flows it will generate, discounted at a rate that reflects the risk inherent to the asset.

General formula:

$$\text{NPV} = \sum_{t=1}^{n} \frac{FCF_t}{(1+r)^t}$$

Where:

- FCF_t = Free Cash Flow in year *t*
- **r** = discount rate
- **n** = projected operating life

11.1.2 Key Steps of the DCF Method Applied to an Oil Asset

11.1.2.1 Modeling of the Production Profile

- Build a gross annual production profile over the remaining life of the contract or field (often until the expiry of the PSC or mining right).
- Contractual allocation:

  - Determine the share due to the State, the operator, and each partner (according to PSC or other regime).
  - Apply the concepts of equity oil, profit oil, and cost oil as per the contract terms.

11.1.2.2 Estimation of Gross Revenues

- Multiply marketable volumes by an assumed oil price.
- Consider several scenarios:

  - Reference scenario (e.g., fixed price of $75/bbl).
  - Optimistic / pessimistic scenarios (±10–20%).

- Include any additional revenues: associated gas, condensates, carbon credits.

11.1.2.3 Deduction of Operating and Investment Costs

a. **OPEX (Operating Expenditures)**

- Recurring costs linked to production: labor, facility maintenance, security, transport, etc.
- Adjust according to asset type (onshore vs offshore, maturity, logistical remoteness).

b. **CAPEX (Capital Expenditures)**

- Investments in equipment, additional drilling, platform upgrades.
- Include abandonment costs (decommissioning, site rehabilitation).

c. **G&A (General and Administrative Costs)**

- Overheads, head office charges, external services.

11.1.2.4 Consideration of Taxation and the Contractual Framework

- **Royalties**: percentage of gross production.
- **Profit taxation**: corporate tax, hydrocarbon tax.
- **Production sharing**: per the PSC model, allocation between cost oil and profit oil, recovery thresholds.
- **Application of specific clauses**: R-factor, sliding scales, bonuses, etc.

11.1.2.5 Calculation of Net Available Cash Flows (FCF)

FCF = Net Revenues – OPEX – CAPEX – Taxes & Royalties

These flows must be estimated annually over the asset's remaining life (often 10–20 years for a mature field).

11.1.3 Choice of Discount Rate

The rate r reflects the Weighted Average Cost of Capital (WACC) adjusted for country risk, technical risk, and contractual risk.

- Stable onshore: 8–10%

- Offshore or high-risk country: 12–15%
- High-risk project (early development, unstable fiscal regime): up to 20%

The choice of discount rate has a significant impact on NPV. A sensitivity analysis is recommended on this parameter.

11.1.4 Final Result: Project or Equity NPV

- The discounted sum of net cash flows over the projected period gives the Net Present Value (NPV) of the field for the production period considered.
- This value can then be weighted by the equity share (e.g., 15% for a minority partner).
- It is possible to add or deduct:

o Environmental liabilities.

o Entry bonuses, transaction costs.

o Upside potential (extensions, development of new reservoirs).

11.1.5 Advantages of the DCF Method

- Industry reference method for producing assets.
- Allows consideration of:

o Timing of cash flows (time value).

o Actual contractual and fiscal structure.

o Technical performance of the field.

11.1.6 Limitations and Precautions

- Sensitive to the quality of data (volumes, prices, costs).
- Risk of error in long-term assumptions (especially oil prices).
- Must be updated regularly in line with changes in production, costs, or regulation.
- Does not value well undeveloped upside potential (un-drilled areas, non-monetized gas).

11.1.7 Practical Recommendations

- Always build a DCF model in Excel or specialized software.
- Include low / base / high scenarios.
- Perform multi-variable sensitivity analysis.
- Ensure production profile is consistent with available reserves.
- Integrate an abandonment plan to account for final costs

11.2 Reserves-Based Valuation

11.2.1 Definition and General Principle

The reserves-based valuation method estimates the economic value of an oil asset by applying a unit value ($/bbl or $/boe) to its marketable reserves volumes. These unit values depend on the geological and technical maturity of the reserves as well as their expected profitability.

This method is often used:

- For a quick first estimate of a field's value.
- In asset transactions (farm-outs, mergers, acquisitions).
- As a complement to a DCF valuation, to serve as a low or high reference point.

11.2.2 Reserves Classification

SPE-PRMS Reference System

Reserves are classified into three main categories depending on their degree of certainty:

- **P1 – Proven:** recoverable with reasonable certainty (>90%), and technically and economically producible under current conditions.
- **P2 – Probable:** additional reserves less certain than P1 but more certain than P3 (>50% probability). Often linked to lateral extensions, future drilling, or new technologies.
- **P3 – Possible:** speculative reserves or those tied to uncertain technical scenarios (>10% probability). Generally excluded from financial cash-flow models.

The combined volumes P1 + P2 are called "2P reserves", and they represent the standard basis for economic valuation of an asset.

11.2.3 Indicative Valuation Grid

Indicative values commonly applied in the industry are as follows:

Type of Reserves	Indicative Value ($/bbl)	Conditions of Application
Proven (P1)	$6 – $12	Producing fields, existing infrastructure
Probable (P2)	$3 – $7	Fields under development, probable extensions
Possible (P3)	$0.5 – $2	Exploration prospects, uncertain potential

Adjustments must be made depending on:

- Geographical and logistical context (onshore, offshore, depth).
- Level of infrastructure integration (processing units, pipelines).
- Regulatory, fiscal, or contractual risks.
- Oil price environment.

11.2.4 Calculation Methodology

Typical steps:

1. Identify equity reserves volumes – determine the company's net entitlement share of reserves after applying contractual terms (e.g., Production Sharing Contract, concession rules, royalty/tax systems).

2. Apply unit values to each reserve category – assign a $/bbl valuation to P1 (proven), P2 (probable), and P3 (possible) reserves, reflecting their commerciality, technical certainty, and development status.

3. Aggregate results – sum the values across categories to obtain a gross estimate of asset value.

Example:

A company holds a 40% participating interest in an offshore block. After contractual allocations (royalties, state share, and partner interests), its net entitlement (equity share) of reserves is as follows:

- P1 (Proven): 15 million barrels
- P2 (Probable): 10 million barrels
- P3 (Possible): 5 million barrels

Industry benchmarks (based on comparable transactions in the region) suggest the following indicative unit values:

- P1 (Proven): $10/bbl (reflecting high certainty, producing status, and existing infrastructure).
- P2 (Probable): $5/bbl (moderate certainty, requiring additional wells and investment).
- P3 (Possible): $1/bbl (low certainty, dependent on exploration success and new technology).

Valuation calculation:

- Equity P1 reserves: 15 million bbl × $10/bbl = $150 million
- Equity P2 reserves: 10 million bbl × $5/bbl = $50 million
- Equity P3 reserves: 5 million bbl × $1/bbl = $5 million

Total estimated value = $205 million

Interpretation of results:

- The majority of value (73%) comes from P1 reserves, which are producing and therefore highly reliable.
- P2 reserves add upside potential but depend on further drilling and CAPEX.
- P3 reserves represent speculative value only and should be treated cautiously in negotiations.

In practice, companies often apply risk factors or discount coefficients to P2 and P3 reserves (e.g., valuing P2 at 50–60% of P1, P3 at 10–20% of P1) to reflect uncertainty. For this reason, while the gross value is shown as $205 million, a risk-adjusted value might be closer to $175–185 million.

This approach provides a quick and transparent estimate, often used in farm-in/farm-out negotiations or as a cross-check against DCF-based valuations.

Variants:

- Some companies apply a risk or discount factor to P2/P3 reserves.
- In complex cases, P2 may be valued at 50–60% of P1 value.

11.2.5 Advantages of the Method

- Simple to apply: does not require detailed financial modeling.
- Allows quick estimates from certification reports or RNS reports.

- Widely used by market participants to establish ballpark values.
-

11.2.6 Limitations and precautions

- Does not account for the time value of money (no discounting of cash flows).
- Does not consider specific operating costs (OPEX, CAPEX, taxation).
- Ignores production profile (plateau, decline rate).
- Can over- or undervalue fields if economic conditions vary significantly.
- Cannot be used alone to validate an investment decision or major transaction.

11.2.7 Best Practices and Recommendations

- Always use certified reserves (external audit SPE/PRMS).
- Adjust unit prices to reflect field location, maturity, and profitability.
- Avoid applying value to P3 reserves without a clear development plan.
- Use this method in conjunction with DCF to cross-check results.
- In transactional contexts, compare valuation with market benchmarks (comparable transactions in the region or for analogous fields).

11.3 Valuation by Multiples

11.3.1 Definition and Principle

The multiples valuation method estimates the value of an oil asset by comparing it to other similar assets whose value is already known—either through recent transactions or through the stock market. The idea is to apply a multiplier coefficient to a key performance indicator (production, reserves, or earnings) in order to derive a market-based valuation. This is a comparative approach built on the principle that assets with similar economic and technical characteristics should command similar valuation ratios. Unlike the DCF method, which models detailed future cash flows, multiples provide a snapshot of value derived from observed market benchmarks.

This method is frequently used in:

- Preliminary asset screenings to establish order-of-magnitude valuations.
- Farm-in/farm-out negotiations where quick comparisons are required.
- Cross-checking results obtained through more detailed valuation methods such as DCF or reserves-based models.

11.3.2 Commonly Used Reference Indicators

The three multiples most frequently applied in the oil and gas industry are:

11.3.2.1 \$/barrel per day of production (b/d)

- Gross daily production is multiplied by a benchmark multiple expressed in \$/b/d.

- Formula:

Value = Daily Production × Multiple ($/b/d)

- Indicative ranges of multiples:

  - Mature onshore assets: $20,000 – $35,000 /b/d
  - Mature offshore assets: $30,000 – $50,000 /b/d
  - High-potential or stable-production assets: $50,000 – $70,000 /b/d

- Contextual interpretation:

These multiples are influenced by factors such as field decline rates, reservoir life, fiscal terms, and oil price expectations. For example, a mature onshore field nearing depletion will trade at the low end of the range, while a producing offshore field with stable output, low costs, and long remaining life may attract higher multiples.

11.3.2.2 EV / EBITDA (Enterprise Value / Earnings Before Interest, Taxes, Depreciation, and Amortization)

- Compares a company's Enterprise Value (EV) to its operating profitability.
- Primarily used for listed companies or assets with stable, predictable cash flows.
- Indicative multiples:

  - Junior exploration & production companies: 3x – 6x EBITDA
  - Integrated companies or majors: 6x – 10x EBITDA

- Contextual interpretation:

EV/EBITDA is especially useful in valuing corporate transactions, as it provides a normalized ratio for comparing companies of different sizes. However, EBITDA multiples tend to vary with oil price cycles: they compress in downturns and expand during commodity price booms.

11.3.2.3 $/barrel of reserves (P1 or 2P)

- A multiple is applied to the quantity of reserves (Proven – P1, or Proven + Probable – 2P).
- Similar to the reserves-based valuation method, but values are drawn from market evidence (comparable deals) rather than theoretical assumptions.
- Typical use: quick valuation of undeveloped or early-stage assets, where production data may not yet be available.

11.3.3 Practical Application: Production Multiple ($/b/d)

This is the most commonly used multiple for quick valuations or in transactions involving producing fields.

The $/b/d production multiple is the most common benchmark for quick valuations, especially for producing assets that have a track record of stable output.

Example of application :

- Current gross production : 4,100 b/d
- Benchmark multiple (mature offshore asset): $40,000 /b/d

Gross field value = 4,100 × $40,000 = $164 million

If a partner holds 55% equity, its net share is:

164 × 0.55 = $90.2 million

This type of multiple does not take into account:

- Remaining field life.

- Future CAPEX or OPEX requirements.
- Fiscal or contractual terms.
- Technical or geological risks.

As such, it should be used only as an indicative metric and always cross-checked against DCF or reserves-based valuations.

11.3.4 Sources of Multiples

a. Comparable Transactions

- Recent asset acquisitions in Africa, the Middle East, or Latin America.
- Public announcements of equity sales or farm-outs (e.g., Wood Mackenzie, Rystad, HIS).

b. **Stock Market Benchmarks**

- Valuations of listed junior companies (TSX, AIM, Oslo Børs).
- Multiples derived from their market capitalization relative to reserves or production.

c. **Internal or Third-Party Benchmarks**

- Internal company or investment bank reports.
- Due diligence carried out by specialized consultancies.

11.3.5 Advantages of the Method

- Fast and simple to apply.
- Requires limited confidential data.

- Useful for initial screening, negotiations, or comparing options.
- Well-suited to producing fields with stable, predictable output.

11.3.6 Limitations and Precautions

- A high-level approach: does not reflect contractual terms or net profitability.
- Risk of misvaluation if comparables are poorly chosen.
- Multiples do not capture future CAPEX, fiscal burdens, or remaining field life.
- Less reliable for geologically heterogeneous assets or those with steep production decline.

11.3.7 Best Practices and Recommendations

- Always justify the chosen multiple: geography, maturity, production profile.
- Avoid relying on a single multiple—test a range (low/base/high).
- Do not base strategic decisions solely on this method.
- Use in combination with DCF or reserves-based valuation to validate results.

11.4 Complementary Methods of Economic Evaluation

Definition

The historical cost or replacement cost method values an asset based on what it would cost to rebuild or replicate it today. This approach considers:

- The cost of existing infrastructure (platforms, wells, processing units).
- Cumulative expenditures already incurred (historical CAPEX).
- Depreciation due to wear, technological obsolescence, or remaining useful life.

This method is widely used in insurance assessments, litigation, or public accounting, where the physical and financial footprint of past investments needs to be recognized.

Simplified formula

Current Value=New Replacement Cost−Technical or Economic Depreciation

Key considerations

- **New replacement cost** represents the estimated expenditure required to construct an identical or equivalent asset with current prices, technologies, and standards.
- **Depreciation** accounts for the reduction in value due to age, physical degradation, or reduced productive capacity compared to a new asset.

It is important to note that, while this method is straightforward and based on tangible data, it does not capture the economic value of future production, reservoir quality, or profitability**.** For this reason, it is considered **a** static and partial approach**,** best suited to specific legal, financial, or regulatory contexts rather than investment decisions.

Use cases

- Insurance or compensation procedures.
- Commercial litigation.

- Public accounting or valuation of assets in nationalization contexts.
- Valuation of pre-production assets with no operational cash flows.

Advantages

- Based on tangible, verifiable data.
- Useful when financial flows are unavailable.

Limitations

- Does not reflect actual economic value.
- Ignores future profitability, subsurface quality, or remaining life.
- Static approach, poorly suited to mature fields.

11.4.2 Comparable Sales Method

This method compares the asset being valued with past transactions involving similar assets in terms of:

- Production size.
- Reserves.
- Location (offshore/onshore, country, logistics).
- Fiscal or contractual framework.

Application

- Analysis of observed transaction prices ($/bbl, $/b/d, EV/EBITDA).
- Adjust valuations for structural differences.

Use cases

- Negotiation of equity sales.
- Due diligence for farm-in or equity entry.
- Comparison with competing offers.

Advantages

- Provides a credible market value.
- Widely used in real transactions, often by M&A firms.

Limitations

- Transactions are not always public.
- Market conditions can change quickly (oil-price volatility).
- Risk of misfit if the asset being valued has unique characteristics.

11.4.3 Strategic (or "Optional") Valuation

Definition

A qualitative method aimed at valuing synergies or strategic advantages that an asset may offer to a specific buyer:

- Logistical integration (proximity to other fields, shared infrastructure).
- Fiscal benefits (profit consolidation, depreciation advantages).
- Integrated development with adjacent blocks or networks.

Example

- A small field adjacent to a terminal owned by the acquirer could hold more value for them than for an independent buyer.
- An acquirer may avoid processing or transport costs through integration.

Advantages

- Can justify a purchase price above the **DCF** valuation (strategic premium).
- Useful in asset consolidation contexts.

Limitations

- Highly subjective.
- Only applies to specific acquirers.
- Requires strong knowledge of the buyer's portfolio.

11.4.4 Real Options Valuation (ROV)

ROV is an advanced financial approach that values future managerial flexibility, such as:

- The right, but not the obligation, to drill a well.
- The option to delay development.
- The opportunity to exploit geological upside in case of exploratory success.

It is inspired by financial option-pricing models (e.g., Black-Scholes, binomial trees), treating future investment decisions as embedded financial options within the asset.

Use cases

- Valuation of exploratory assets with high uncertainty.
- Modular or scalable projects.

- Portfolios with multiple deferred decisions.

Advantages

- Captures the value of flexibility (waiting, expansion, phasing).
- Robust in project finance contexts.

Limitations

- High mathematical complexity.
- Requires detailed data and numerous assumptions.
- Rarely used as a standalone method for producing assets.

11.4.5 Comparative Synthesis of Complementary Methods

Method	Applicability	Specificity	Relevance for Mature Fields
Historical cost / replacement	Technical, litigation	Infrastructure, compensation	Low
Comparable sales	M&A, secondary market	Observable market prices	Medium to high
Strategic valuation	Synergies, integration	Buyer-specific advantage	Medium
Real options	Exploratory, flexible	Probabilistic modeling	Low

Complementary methods serve as reinforcement, verification, or strategic framing tools, but they do not replace the fundamental approaches (DCF, reserves-based valuation, multiples).

They are particularly useful :

- When financial data are incomplete or unavailable.
- In contexts of litigation, compensation, or internal valuation.
- When an asset presents specific advantages not captured by cash-flow 161odelling.

11.5 Best Practice Recommendations for the Evaluation of Petroleum Assets

The evaluation of a petroleum asset—whether in production, development, or exploration—must be carried out within a rigorous, structured process, adapted to the asset's technical, contractual, geographic, and economic specificities. A solid evaluation depends as much on the quality of the methods used as on the intelligence with which they are applied.

11.5.2 Combine Valuation Methods

Principle

No single method, when used in isolation, can provide a complete picture of an asset's value. It is therefore recommended to cross-check results from several approaches in order to:

- Validate the consistency of estimates.
- Identify abnormal gaps or structural inconsistencies.
- Present a credible value range (low, median, high).

Best practice

- **DCF:** reference method → base of the valuation.
- **Reserves-based approach:** physical verification of the substance of the asset.
- **Multiples :** market comparison point.
- **Complementary methods:** contextualization or strategic adjustment.

11.5.3 Use Reliable Technical and Contractual Data

A robust evaluation relies on high-quality data:

- Historical and forecast production profiles consistent with reality.
- Reserve certifications in line with PRMS standards (P1, P2, P3).
- Petroleum contract structure (PSC, concession, royalty/tax) and its fiscal/economic implications.
- Detailed OPEX and CAPEX costs, with expected trends.
- Oil price assumptions based on credible market scenarios.

Therefore, as a best practice:

- Work with data from the asset's technical custodian (operator or petroleum authority).
- Use independent auditors (e.g., GaffneyCline, DeGolyer & MacNaughton, NSAI, etc.) for volumes.
- Update data whenever production profiles or economic conditions change.

11.5.3 Integrate Sensitivity Analysis

The results of an economic evaluation are highly sensitive to the assumptions used (prices, costs, fiscal terms, discount rates, etc.). It is therefore essential to conduct a multi-parameter sensitivity analysis in order to:

- Identify key risk factors.
- Highlight breakeven thresholds.
- Build a robust case for investors, partners, or regulators.

Examples of variables to test

- Oil price: ±10–20%.
- Discount rate: ±2 points.
- Production plateau variation or accelerated decline.
- OPEX variation: ±15–30%.

11.5.4 Produce Scenario-Based Valuations

Rather than presenting a single point estimate, it is recommended to reason in terms of economic scenarios, reflecting the potential variability of outcomes:

Scenario	Key Assumptions	Expected Result
Low (pessimistic)	Low prices, rapid decline, high costs	Very conservative net value
Medium (realistic)	Central assumptions	Reference estimate
High (optimistic)	Geological upside, high prices	Maximum theoretical valuation

This approach enables more informed decision-making, identification of negotiation margins, and structuring of contractual mechanisms (e.g., earn-outs, performance-based bonuses).

11.5.5 Check Temporal and Contractual Consistency

The value of an asset depends not only on production and reserves, but also on:

- Remaining license term (e.g., PSC expiry, renewal rights).
- Regulatory obligations (local content, decommissioning).
- Restrictive clauses (cost-recovery ceilings, R-factor, marketing limits).

Therefore, as a best practice:

- Always cross-check economic evaluation with the legal and contractual framework of the license.
- Integrate end-of-life costs (abandonment, rehabilitation) into cash flows.
- Verify compliance with local tax laws (exemptions, deferred taxation, etc.).
-

11.5.6 Document the Process and Justify Assumptions

A sound evaluation must be traceable, transparent, and replicable. Accordingly, the valuation report or memo should:

- Clearly explain the assumptions adopted.
- Document the sources used (technical reports, contracts, benchmarks).
- State the limitations of the analysis.

- Provide uncertainty ranges.

This is essential in cases of negotiation, regulatory review, or internal audit.

11.5.7 Adhere to International Standards

It is recommended to align the evaluation process with recognized standards:

- PRMS (Petroleum Resources Management System) : reserve classification.
- SPE guidelines : technical terminology.
- IFRS or US GAAP: accounting-based financial evaluation.
- Technical or financial audit standards for transactions (VDD, QDD).

11.5.8 Seek Independent Third-Party Validation

When the evaluation is used in a transactional, regulatory, or strategic context, it is recommended to:

- Have the model reviewed by an independent specialist consultancy.
- Obtain a fairness opinion report if the asset is being valued in a related-party transaction.
- Request a technical certification of reserves or production data.

Best practices in petroleum asset valuation rest on:

- A rigorous methodological approach.
- Integrity of assumptions.
- Transparency of results.

By applying these principles, evaluation becomes a reliable tool for decision-making, negotiation, and regulation—contributing to the sustainable management of national resources and the protection of stakeholders' interests.

The economic evaluation of petroleum assets is a strategic exercise of critical importance—for States, investors, contractual partners, and regulators alike. In a sector characterized by technical complexity, price volatility, and contractual specificities, a reliable and rigorous estimate of the value of an oil field cannot rely on a single method.

In the particular context of mature fields—subject to constrained investment cycles, limited remaining production life, and declining output—methodological rigor becomes even more decisive. Any decision regarding equity sales, entry of new partners, or contractual renegotiations must be firmly based on an evaluation that is robust, up to date, and aligned with international standards.

Finally, asset valuation is not limited to financial appreciation. It also directly impacts economic sovereignty, resource governance performance, and partner confidence. A well-executed evaluation therefore contributes directly to the quality of strategic decision-making in the petroleum sector.

Key Takeaways

Chapter 11 highlights the strategic importance of the economic evaluation of petroleum assets, which underpins decisions related to transactions, partnerships, and investment planning. It emphasizes that no single method is sufficient on its own and that robust valuation requires the combined use of several approaches.

The Discounted Cash Flow (DCF) method remains the industry's reference tool, providing a detailed view of asset value by discounting projected free cash flows. However, it is sensitive to assumptions such as oil price, production profiles, and discount rates. Reserves-based valuation offers a simpler physical measure, applying indicative values to certified reserves, while the multiples method provides market-oriented benchmarks for negotiations.

The chapter also reviews complementary methods—including historical cost, comparable transactions, strategic valuation, and real options—which are especially useful in specific contexts such as litigation, asset integration, or high-uncertainty projects.

Best practice recommendations stress the need for reliable technical and contractual data, sensitivity and scenario analyses, and strict alignment with international standards (PRMS, IFRS, SPE guidelines). Transparency, documentation, and validation by independent third parties are identified as essential conditions for ensuring credibility.

Ultimately, the chapter concludes that the value of an asset is best presented as a range of outcomes—low, median, and high—rather than as a single figure. This approach reflects both technical realities and market uncertainties, providing decision-makers with a more reliable basis for negotiation, investment, and long-term resource management.

Conclusions

The management of NOPAs constitutes a key strategy in a sector where financial, technical, and environmental challenges are immense. This investment model is built on a philosophy of collaboration and shared responsibility, enabling companies to maximize returns while limiting direct exposure to operational risks. Partnerships established within this framework strengthen the capacity of companies to navigate complex and often volatile environments.

For States, these assets represent a unique opportunity to optimize natural resources while benefiting from the technologies and capital brought by international investors. However, to ensure the success of such partnerships, it is essential to establish robust regulatory frameworks, reinforce transparency, and invest in the development of local expertise. Implementing rigorous monitoring systems, combined with proactive risk management, is critical to protecting national interests and ensuring sustainable exploitation.

Despite their many advantages, NOPAs are not free of risks. Oil-price fluctuations, technical incidents, and environmental pressures demand constant vigilance. Historical examples demonstrate that preparation, cooperation, and innovation are decisive factors in overcoming these challenges.

Ultimately, success in managing NOPAs rests on a delicate balance between strategic collaboration, adoption of industry best practices, and adherence to environmental and social standards. This model reflects a modern and responsible vision of petroleum exploitation, where economic benefits can be aligned with the imperatives of sustainability and development. Both public and private stakeholders stand to gain by deepening their understanding of these dynamics in order to fully seize the opportunities they present.

To Explore the Topics in Greater Depth

Chapter 1: Understanding NOPAs

1. BP (2022) *Statistical Review of World Energy 2022*. BP.

2. Extractive Industries Transparency Initiative (EITI) (2019) *Transparency in Oil and Gas Contracts*. EITI Secretariat.

3. International Energy Agency (IEA) (2020) *Oil Market Report 2020*. IEA.

4. Johnston, D. (1994) *International Petroleum Fiscal Systems and Production Sharing Contracts*. PennWell.

5. Kaiser, M.J. (2018) *Decommissioning Forecasting in the Offshore Oil and Gas Industry*. Elsevier.

6. Likosky, M. (2009) *Contracting and Regulatory Issues in the Oil and Gas and Mining Sectors*. Routledge.

7. Radon, J. (2007) 'The ABCs of petroleum contracts: license-concession agreements, joint ventures, and production-sharing agreements', *Transnational Law & Contemporary Problems*, 16(1), pp. 89–120.

8. Tordo, S., Tracy, B.S. and Arfaa, N. (2011) *National Oil Companies and Value Creation*. World Bank.

Chapter 2: Fundamental Principles of Rational Management

1. Auty, R.M. (1993) *Sustaining Development in Mineral Economies: The Resource Curse Thesis*. Routledge.

2. Global Witness (2018) *Corruption Risks in Oil and Gas Revenue Management*. Global Witness.

3. Hartwick, J.M. (1977) 'Intergenerational equity and the investing of rents from exhaustible resources', *American Economic Review*, 67(5), pp. 972–974.

4. Humphreys, M., Sachs, J. and Stiglitz, J.E. (2007) *Escaping the Resource Curse*. Columbia University Press.

5. International Monetary Fund (IMF) (2019) *Taxation in the Oil and Gas Sector*. IMF.

6. International Monetary Fund (IMF) (2021) *Fiscal Regimes for Extractive Industries*. IMF.

7. Organization of the Petroleum Exporting Countries (OPEC) (2014) *World Oil Outlook 2014*. OPEC.

8. Smith, J.L. (2014) 'Issues in the valuation of upstream oil and gas investments', *Energy Economics*, 46, pp. 440–452.

9. World Bank (2020) *Managing Oil Revenues for Long-Term Growth*. World Bank.

Chapter 3: Technical Oversight of Non-operated Petroleum Assets

1. Inkpen, A.C. and Moffett, M.H. (2011) *The Global Oil & Gas Industry: Management, Strategy and Finance.* PennWell.

Chapter 4: Financial Oversight and Revenue Optimization

1. Campbell, J.R. and Wood, C.L. (2011) *Fundamentals of Oil & Gas Accounting* (6th ed.). PennWell.

2. Inkpen, A.C. and Moffett, M.H. (2011) *The Global Oil & Gas Industry: Management, Strategy and Finance*. PennWell.

Chapter 5: Cooperation and Communication with Operators

1. Daft, R.L. and Lengel, R.H. (1986) 'Organizational information requirements, media richness and structural design', *Management Science*, 32(5), pp. 554–571.

2. Hilyard, J. (2012) *The Oil & Gas Industry: A Nontechnical Guide*. PennWell.

3. Inkpen, A.C. and Moffett, M.H. (2011) *The Global Oil & Gas Industry: Management, Strategy and Finance*. PennWell.

4. Luthans, F. and Doh, J.P. (2018) *International Management: Culture, Strategy, and Behavior* (10th ed.). McGraw-Hill.

5. Moran, R.T., Harris, P.R. and Moran, S.V. (2011) *Managing Cultural Differences: Global Leadership Strategies for Cross-Cultural Business Success* (8th ed.). Routledge.

6. Rubin, J. (2009) *Why Your World Is About to Get a Whole Lot Smaller: Oil and the End of Globalization*. Random House Canada.

Chapter 6: Strategies for Continuous Improvement

1. Calixto, E. (2016) *Gas and Oil Reliability Engineering* (2nd ed.). Gulf Professional Publishing/Elsevier.

2. Campbell, J.D. and Reyes-Picknell, J.V. (2015) *Uptime: Strategies for Excellence in Maintenance Management* (3rd ed.). CRC Press.

3. Campbell, J.D., Jardine, A.K.S. and McGlynn, J. (2024) *Asset Management Excellence: Optimizing Equipment Life-Cycle Decisions* (3rd ed.). CRC Press.

4. Chowdhury, S. (2016) *Optimization and Business Improvement Studies in Upstream Oil and Gas Industry*. Wiley.

5. Deighton, M.G. (2016) *Facility Integrity Management: Effective Principles and Practices for the Oil, Gas and Petrochemical Industries* (2nd ed.). Gulf Professional Publishing/Elsevier.

6. El-Reedy, M.A. (2022) *Asset Integrity Management for Offshore and Onshore Structures*. Gulf Professional Publishing/Elsevier.

Chapter 7: Basic Practical Rules for Managing an Oil Asset

1. Center for Chemical Process Safety (CCPS) (2017) *Guidelines for Asset Integrity Management*. Wiley/AIChE.

2. Friedenthal, J. and Reeve, D. (2016) *Performance Management for the Oil, Gas, and Process Industries*. Elsevier.

3. Inkpen, A.C. and Moffett, M.H. (2011) *The Global Oil & Gas Industry: Management, Strategy and Finance*. PennWell.

4. Speight, J.G. (2015) *Rules of Thumb for Petroleum Engineers*. Wiley.

5. Velmurugan, R.S. and Dhingra, A.K. (2021) *Asset Maintenance Management in Industry: A Comprehensive Guide to Strategies, Practices and Benchmarking*. Springer.

Chapter 8: Contractual Dynamics and Rent Sharing

1. Bindemann, K. (1999) *Production-Sharing Agreements: An Economic Analysis*. Oxford Institute for Energy Studies.

2. Darmois, G. (2014) *Sharing the Oil Rent: Current Situation and Good Practice*. Éditions Technip.

3. Johnston, D. (1994) *International Petroleum Fiscal Systems and Production Sharing Contracts*. PennWell.

Chapter 9: Core Financial Indicators

1. Hey, R.B. (2017) *Performance Management for the Oil, Gas, and Process Industries: A Systems Approach.* Elsevier.

2. Inkpen, A.C. and Moffett, M.H. (2011) *The Global Oil & Gas Industry: Management, Strategy and Finance.* PennWell.

3. Johnston, D. (2006) *Introduction to Oil Company Financial Analysis.* PennWell.

4. Tague, J. (2018) *Oil & Gas Performance Analysis: A Practical Guide.* PennWell.

Chapter 10: Integration of Indicators

1. Hey, R.B. (2017) *Performance Management for the Oil, Gas, and Process Industries: A Systems Approach.* Elsevier.

2. Holdaway, K. (2014) *Harness Oil and Gas Big Data with Analytics*. Wiley.

3. Inkpen, A.C. and Moffett, M.H. (2011) *The Global Oil & Gas Industry: Management, Strategy and Finance*. PennWell.

4. Tague, J. (2019) *Oil & Gas Performance Analysis: A Practical Guide for Managers, Engineers and Field Personnel*. PennWell.

Chapter 11: Economic Evaluation of Petroleum Assets

1. Johnston, D. (1994) *International Petroleum Fiscal Systems and Production Sharing Contracts*. PennWell.

2. Kasriel, K. and Wood, D. (2013) *Upstream Petroleum Fiscal and Valuation Modeling in Excel: A Worked Examples Approach*. Wiley.

3. Masseron, J. (1990) *Petroleum Economics: Theory and Practice*. Editions Ophrys.

4. Mun, J. (2005) *Real Options Analysis: Tools and Techniques for Valuing Strategic Investments and Decisions* (2nd ed.). Wiley.

5. Simkins, B. and Simkins, R. (2013) *Energy Finance and Economics: Analysis and Valuation, Risk Management, and the Future of Energy*. Wiley.

GLOSSARY

A

Abandonment of wells: The procedure by which an uneconomic or end-of-life well is permanently shut down and sealed according to strict environmental standards.

Audit of compliance: A verification process ensuring that exploration, development, and production operations comply with applicable standards and regulations.

B

Barrel: A unit of measurement for crude oil, equivalent to approximately 159 liters.

Benchmarking: Systematic comparison of a project's performance with that of similar projects to identify and adopt best practices.

Block (oil block): A geographical area awarded for petroleum exploration and exploitation, often defined through a state or international licensing round.

C

CAPEX (Capital Expenditure): Capital investment expenditures for the development of petroleum infrastructure and equipment (drilling, facilities, pipelines).

Cash flow : A financial indicator representing the net amount of cash generated by a company after deducting all expenses and investments.

Concession (oil concession): An agreement by which a State grants a company exclusive rights to explore and produce hydrocarbons in a specific area for a defined period.

D

Drilling: The technical process of creating a wellbore in the subsurface to access hydrocarbon reservoirs.

E

EITI (Extractive Industries Transparency Initiative): An international initiative promoting transparency in the management and distribution of revenues from natural resource exploitation.

Enhanced Oil Recovery (EOR): Techniques to improve oil recovery in a reservoir by injecting fluids (water, gas, chemicals) or applying heat.

Exploitation (oil exploitation): The process of extracting crude oil from a reservoir for commercialization and refining.

Exploration (oil exploration): All activities and techniques (seismic, geophysics, geology) aimed at locating exploitable hydrocarbon reservoirs.

Exploration license: An authorization granted by a state authority allowing a company to explore designated areas for hydrocarbons.

F

Flaring: The burning of excess natural gas, often criticized for its environmental impact and subject to reduction policies.

G

Geological qualification: A preliminary assessment process of a basin or exploration area to determine hydrocarbon potential, based on seismic surveys, field studies, and geological analyses.

Geosciences: The set of disciplines (geology, geophysics, geochemistry) used to study subsurface formation, structure, and composition, essential for oil exploration.

Governance (petroleum governance): The set of rules, practices, and institutions governing petroleum resource management, ensuring transparency, fairness, and environmental compliance.

H

Hydraulic fracturing: A technique that injects fluids at high pressure into the subsurface to fracture rock and facilitate hydrocarbon extraction, notably in shale formations.

Hydrocarbons: Organic compounds composed of carbon and hydrogen, encompassing crude oil and natural gas, a major global energy source.

Hydrocarbon recovery factor: The percentage of oil or gas extracted relative to the total estimated volume in the reservoir.

I

Injection of fluids: A process used to maintain reservoir pressure or enhance hydrocarbon recovery through the controlled introduction of fluids.

IOGP (International Association of Oil & Gas Producers): An international organization of leading producers that sets standards and promotes knowledge-sharing in the industry.

J

Joint venture: A partnership between several companies to pool expertise, risks, and investments in the development of a petroleum project.

M

Margin (profit margin): The difference between revenues generated by an oil project and all costs associated with its operation and investments.

N

National Oil Company (NOC): A state-controlled company playing a strategic role in managing and developing national petroleum resources.

Net Present Value (NPV): A financial indicator measuring project profitability by discounting future cash flows.

NOPAs : A financial participation in a petroleum project without assuming responsibility for operational management.

O

Offshore: Refers to oil activities conducted at sea, requiring specialized infrastructure (platforms, vessels, subsea pipelines).

Oil operator: A company responsible for the technical and operational management of a petroleum project, including planning, drilling, production, and maintenance.

Oil investment: Capital committed across all phases of a petroleum project, from exploration to production, with particular attention to technical and geopolitical risks.

OPEX (Operating Expenditure): Recurring costs associated with the day-to-day operation of a petroleum project (maintenance, logistics, personnel, etc.).

Optimization of revenues: Strategies and techniques to maximize project profitability through production optimization, cost control, and adaptation to market fluctuations.

P

Petrophysics: The study of the physical properties of reservoir rocks and the fluids they contain, essential for assessing reservoir performance.

Pipeline: Infrastructure for transporting crude oil or gas over long distances, linking production sites to refineries or distribution markets.

Platform (offshore platform): A structure installed at sea to conduct exploration, drilling, and oil production operations.

Production royalties: Payments made by a company operating an oil field to the State or resource owners, calculated based on production volumes or generated revenues.

Production Sharing Contract (PSC): A contractual agreement between a State and a company defining how field production is divided after deduction of operating costs.

Petroleum taxation: The set of taxes, royalties, and levies applied to oil companies on their investments, revenues, and production, defining the economic framework of operations.

Proven reserves: Estimated quantity of oil or gas recoverable with a high degree of certainty, considering economic and technical constraints.

R

Recovery (secondary recovery): Techniques (e.g., water or gas injection) used to maintain reservoir pressure and increase recoverable hydrocarbons after primary production.

Regulatory framework: The set of laws, regulations, and directives governing the exploration, exploitation, and management of hydrocarbons, ensuring safety, environmental protection, and equitable resource distribution.

Reserve level: Classification of a field's reserves (proven, probable, possible) based on the technical and economic certainty of extraction.

Reservoir: A subsurface formation containing recoverable hydrocarbons, characterized by its size, porosity, and permeability.

Reservoir modeling: A numerical and geological process to simulate the future behavior of a reservoir and optimize production.

Risk (financial risk): Uncertainty linked to oil-price fluctuations, operating cost variations, and macroeconomic conditions affecting project profitability.

Royalties: Payments to the State or landowners in exchange for the right to exploit a reservoir, usually calculated based on production or revenues.

S

Safety (oilfield safety): Preventive measures and intervention protocols designed to protect personnel, infrastructure, and the environment from risks associated with oil operations.

Seismic survey: A geophysical exploration method using shock waves to probe subsurface structures and identify potential reservoirs.

Surveillance of non operated petroleum assets: Oversight and evaluation processes implemented by non-operating investors to ensure project performance and profitability.

T

Terrain of exploitation: The geographical zone where an oil company conducts exploration and production activities.

U

Upstream: Refers to exploration and production activities in the petroleum industry, upstream of the value chain.

Utilization of resources (optimized): An integrated approach to maximize hydrocarbon recovery while minimizing environmental impact and improving economic performance.

V

Value of petroleum assets: The overall economic estimate of resources, infrastructure, and exploitation rights linked to a project or investment portfolio in the oil sector.

Volatility of oil prices: Rapid and unpredictable fluctuations in international oil markets, directly impacting project profitability.

W

Waterflooding: A secondary recovery technique involving water injection into the reservoir to maintain pressure and mobilize hydrocarbons.

Well log: A set of technical and geological data collected during drilling, used to assess formations encountered and guide operational decisions.

Well testing: A set of procedures to assess the productivity and characteristics of a well, including pressure, flow rate, and fluid composition.

X

X-Factor: A term describing unpredictable or non-quantifiable variables that may affect project profitability and performance, such as geopolitical risks or regulatory changes.

Y

Year-on-Year (YoY): An indicator comparing a project's or company's performance over one year, useful for analyzing sector trends and evolution.

Yield: The ratio between the amount of oil recovered and the theoretical or expected capacity of a reservoir, often expressed as a percentage.

Z

Zero Flaring Policy: A policy adopted by some companies aiming to eliminate routine flaring of natural gas to reduce greenhouse-gas emissions and operational carbon footprint.

Zone (drainage area): The subsurface zone from which fluids (oil, gas, water) migrate toward a production well, influencing reservoir dynamics.

Zone (fractured zone): A portion of a reservoir where the rock contains natural fractures, facilitating hydrocarbon flow and influencing extraction strategy.

Table of contents

Structures éditoriales du groupe L'Harmattan

L'Harmattan Italie
Via degli Artisti, 15
10124 Torino
harmattan.italia@gmail.com

L'Harmattan Hongrie
Kossuth l. u. 14-16.
1053 Budapest
harmattan@harmattan.hu

L'Harmattan Sénégal
10 VDN SICAP AMITIE 3
BP 45034 Dakar-Fann
direction@senharmattan.com

L'Harmattan Cameroun
Tsinga/Derrière polyclinique FECAFOOT
BP 11486 Yaoundé
Jacques.koukam@harmattan.fr

L'Harmattan Burkina Faso
Villa 281, secteur 23 623, rue 14.32
Ouagadougou
zagbie@outlook.fr

L'Harmattan Guinée
Almamya, rue KA 028 OKB Agency
BP 3470 Conakry
harmattanguinee@yahoo.fr

L'Harmattan RDC
185, avenue Nyangwe
Commune de Lingwala – Kinshasa
matangilamusadila@yahoo.fr

L'Harmattan Congo
219, avenue Nelson Mandela
BP 2874 Brazzaville
deunovmikhael@gmail.com

L'Harmattan Mali
Immeuble Mgr Jean Marie Cissé
Aci Hamadallaye
BP 145 Bamako
syllaka@yahoo.fr

L'Harmattan Togo
Djidjole – Maison Amela
face EPP BATOME
02 BP 20680 Lomé
didieramela11@gmail.com

L'Harmattan Côte d'Ivoire
Résidence Karl – Cité des Arts
Abidjan-Cocody
03 BP 1588 Abidjan
harmattan.ivoire@gmail.com

L'Harmattan Bénin
Rue 604, (Immeuble à l'entrée
de la rue UCAO- Cite de Saint-Jean)
Cotonou
stepcorpus@gmail.com

Nos librairies en France

Librairie internationale
16, rue des Écoles
75005 Paris
librairie.internationale@harmattan.fr
01 40 46 79 11
www.librairieharmattan.com

Librairie des savoirs
21, rue des Écoles
75005 Paris
librairie.sh@harmattan.fr
01 46 34 13 71
www.librairieharmattansh.com

www.ingramcontent.com/pod-product-compliance
Lightning Source LLC
Chambersburg PA
CBHW071445230926
51846CB00001B/351